Guglielmo Lanzani

The Photophysics behind Photovoltaics and Photonics

Related Titles

Ware, W. R.

Introduction to Organic Molecular Photophysics

2011

ISBN: 978-0-471-63397-6

Rafailov, E. U., Cataluna, M. A., Avrutin, E. A.

Ultrafast Lasers Based on Quantum Dot Structures

Physics and Devices

2011

ISBN: 978-3-527-40928-0

Krüger, A.

Carbon Materials and Nanotechnology

2010

ISBN: 978-3-527-31803-2

Radons, G., Rumpf, B., Schuster, H. G. (eds.)

Nonlinear Dynamics of Nanosystems

2010

ISBN: 978-3-527-40791-0

Hofmann, P.

Solid State Physics

An Introduction

2008

ISBN: 978-3-527-40861-0

Siebert, F., Hildebrandt, P.

Vibrational Spectroscopy in Life Science

2008

ISBN: 978-3-527-40506-0

Meschede, D.

Optics, Light and Lasers

The Practical Approach to Modern Aspects of Photonics and Laser Physics

2007

ISBN: 978-3-527-40628-9

Guglielmo Lanzani

The Photophysics behind Photovoltaics and Photonics

WILEY-VCH

WILEY-VCH Verlag GmbH & Co. KGaA

The Author

Prof. Guglielmo Lanzani
IFN - Dip. di Fisica
Politecnico di Milano
Milano, Italy
guglielmo.lanzani@iit.it

Library of Congress Card No.: applied for

British Library Cataloguing-in-Publication Data
A catalogue record for this book is available from the British Library.

Bibliographic information published by the Deutsche Nationalbibliothek
The Deutsche Nationalbibliothek lists this publication in the Deutsche Nationalbibliografie; detailed bibliographic data are available on the Internet at <http://dnb.d-nb.de>.

Typesetting Laserwords Private Limited, Chennai, India
Printing and Binding Markono Print Media Pte Ltd, Singapore
Cover Design Adam-Design, Weinheim
Printed in Singapore
Printed on acid-free paper

Print ISBN: 978-3-527-41054-5
ePDF ISBN: 978-3-527-64515-2
oBook ISBN: 978-3-527-64513-8
ePub ISBN: 978-3-527-64514-5
mobi ISBN: 978-3-527-64516-9

Contents

1
Introduction

During a conference, in May 2003, the Nobel laureate Richard Smalley proposed a list of humanity's top 10 problems for the next 50 years. The list was (i) energy, (ii) water, (iii) food, (iv) environment, (v) poverty, (vi) terrorism and war, (vii) disease, (viii) education, (ix) democracy, and (x) population. This is a quite reasonable list, easy to agree with. The most striking point is that there is essentially one and only one problem, because each issue in the list is tightly correlated to all the others. Take anyone of the problems, and try to find connections to the others. For instance, let us take no. (iii) "food." Energy is needed to produce and distribute food; quality and quantity of water is also related to food production (both for agriculture and farming); environment dictates the quality of food (pollution) and is affected by food production techniques; poverty has much to do with lack of food; terrorism and war brings poverty and famine, and vice versa; diseases might be a consequence of food shortage or food deterioration (food conservation requires energy and education); education affects the way citizens choose and dispose of food; democracy has to do with wealth distribution, including food; and population growth poses an ever rising demand for food. There is one single problem; we can name it *sustainable development.*

In 1990, the world's total energy consumption (as primary power) was about 12 TW, with 5.3 billion people (US Census Bureau, International Database). In 2050, it is expected to be 28 TW, with a population of 8–10 billion people. Although these numbers are subject to large fluctuations, they point out a solid truth: energy demand will keep growing.

Another well-founded evidence is that fossil fuels, oil, natural gas, and coal are doomed to end. In particular, many share the view that the end of the oil era is approaching. The end itself can hardly be questioned, but when it will happen is a matter of debate. The fuel consumption curve is called *Hubbert plot.* Hubbert was a geophysicist working for Shell Oil Company back in the 1960s. His plot describes production of oil (or any other fossil fuel) versus time. It is obtained as a derivative of the logistic curve, $Q(t) = \frac{Q_\infty}{1+e^{A(\tau_m - t)}}$, which describes self-limiting growth (for instance, population growth when resources are proportionally reduced). The Hubbert plot $P_H(t) = \frac{dQ}{dt} = AQ_\infty \frac{e^{A(\tau_m - t)}}{[1+e^{A(\tau_m - t)}]^2}$ is very similar to a Gaussian plot. The rising slope is when new oil wells are localized. The slope slows down when

The Photophysics behind Photovoltaics and Photonics, First Edition. Guglielmo Lanzani.

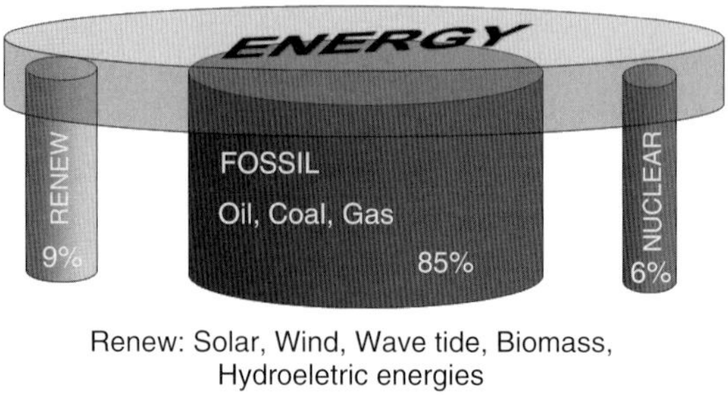

Figure 1.1 The breakdown in energy sources of the world energy supply (2005 data).

discovering new resources gets difficult and the extraction cost increases, to reach the peak. The peak is when half of the available reserves have been used. Decay is expected at the same rate of increase, giving rise to the characteristic symmetric bell shape. In real world, there is no reason for a symmetric shape, because many economic variables might change it, usually giving a slower decay tail. Fitting with true data, the increasing side allows "predictions" on peak and decay, which in 1970 for US oil turned out quite accurate. The peak is when half of the total reserve has been extracted. In 2011, most predictions place the world oil peak around 2040. This is particularly dramatic since oil itself takes about 30% of the energy balance, and including other fossil fuels (gas and coal), the figure approaches 90% (Figure 1.1).

The question of alternatives to oil is thus a hot one for this century. Nuclear energy might be an alternative, provided that hard environmental threats such as the storage of waste, the supply of raw material, and delicate political issues linked to military applications can be kept under control. The discussion among experts about which energy source can be cheaper is unfortunately never a scientific one, but most of the time it is a battle of religion. Each side struggles to support its own business (that could provide jobs, money, and power to the proponents). Having an objective opinion is difficult. While nuclear energy can be a choice for some of the countries with proper geopolitical conditions, it is not a universal solution and most certainly not the stable solution, because it is not renewable. Coal is another intermediate alternative, with its own problems. Coal is available in huge stocks, yet limited, but its use poses a considerable environmental challenge: burning coal produces large amount of carbon dioxide and other green house gases. It is curious to note that the negative effect on the environment of burning coal is seen as a social problem since long time ago. In 1273, King Edward I so spoke to his parliament: *Be it known to all within the sound of my voice, whosoever shall be found guilty of the burning of coal (in London) shall suffer the loss of his head.* This is sometime considered the first antipollution law and, needless to say, it was a very unsuccessful law.

The green house effect is seen as one of the worst challenge to planet stability, so carbon capture and sequestration (CCS) is an issue to be solved if coal has to be used. There is a "low carbon roadmap," considering the connection between gross domestic product (GDP) and carbon emission, stated in the equation

$GDP = C \times (E/C) \times (GDP/E)$, which regards the increase in economic productivity (GDP) without increasing carbon emission (C). The second term, E/C, concerns the amount of energy produced per carbon emitted, while the third term GDP/E regards wealth over energy consumption. Both should be enhanced, and we need a research and development program aimed at increasing GDP in a carbon-constrained world. The G7 is moderately energy efficient, with about \$30 GDP per capita and an energy efficiency of \$160 of GDP per billion joule. The United States and Canada are the less efficient in the group. As an example, Russia, Saudi Arabia, and Iran are rather inefficient and low productive (with GDP per capita below \$15 000), while in Asia, Philippines and Bangladesh are highly energy efficient but with the lowest productivity and GDP per capita within the top 40 economy (for GDP), with GDP per capita below or well below \$5000. The case of the United States is indeed instructive. This rough trend tells us that where large energy stocks are available there is little care for efficiency, in spite of technology and wealth. Nations of the world have a long way ahead to reach sustainable efficiency. Meanwhile, the oil age is near its end. There are skeptics, who do not believe this, because they remind the burst in 1970 that led to nothing. At that time, we had an oil price crisis due to the prediction of extinguishing oil reserves at around 2000. Indeed, nothing happened because new resources were found and extraction technology improved. It is very hard, however, that the same could happen again, for many reasons. The extraction technologies might still improve, yet at exponential growth in the cost. New reserves, based on the present scanning and searching technology that allows looking at the planet as a transparent sphere, simply do not exist.

In summary, three reasons for rethinking our energy strategy are shortage, environment, and demand.

Scientists have an important role in the process of innovating energy strategy. First, break through in basic science is needed. We need new principles for energy harvesting and conversion, new materials, and new concept for devices. Nanoscience and bioscience stay on the forefront and offer the best chance for this innovation. Among the many possible solutions, the more extended use of solar energy plays a big role.

Solar radiation is a huge source of energy, about 170 PW. The average radiation intensity is 1000 Wm^{-2}, which reduces to 10 in cloudy and polluted town atmosphere. Yet, it is estimated that covering 0.16% of the earth's land with 10% efficient solar panels would be enough to produce 20 TW, the expected planet demand around 2050. Solar radiation is the only known, safe, and reliable energy source from a nuclear fusion reactor, the sun. It is a clean energy source with low environmental impact, and it is renewable. Even if shortage of other sources is not an imminent threat, solar energy has many other appealing features to deserve attention from scientist and engineers. As writer Ian McEwan puts it, in his enjoyable novel "Solar," *the stone age didn't end because of a shortage of stones.*

Solar energy can be converted into thermal energy (solar thermodynamic), heating up high thermal capacity materials; it can be used to produce hydrogen and store it away; or it can be directly transformed into electrical energy. Solar thermodynamics is better suited for large plants; hydrogen production may be the future solution but it is still quite immature and difficult to be implemented. Solar photovoltaics are suited for portable energy sources and for local use in domestic or small, remote areas. This is expected to be the next revolution in energy use.

The beginning of photovoltaics is attributed to Alexandre Edmond Becquerel, who discovered a physical phenomenon allowing light–electricity conversion. Willoughby Smith discovered the photovoltaic effect in selenium in 1873. In 1876, with his student R. E. Day, William G. Adams discovered that illuminating a junction between selenium and platinum also has a photovoltaic effect. These two discoveries were the foundation for the first selenium solar cell construction, which was built in 1877. Photovoltaics remained a curiosity till silicon came into play. Early silicon solar cells date back to 1940, but the breakthrough occurred at Bell labs, when Gerald Pearson, a physicist, built, apparently involuntarily, a silicon solar cell with efficiency much higher than that of selenium cells. Improved by two other scientists at Bell – Darryl Chapin and Calvin Fuller – the Bell silicon solar cells could work with 6% efficiency on a sunny day.

This immediately attracted the interest of engineers of the most powerful armies of the time, the United States and Soviet Union, which well understood that such photovoltaic cells were best fitted for powering satellites in the cold war of space race. Curious enough, the first good customer for photovoltaic energy is the oil extraction industry. Photovoltaic cells were used on oil-drilling rigs in the Mexican Gulf for powering safety lights. Perhaps the more intriguing use is, however, in remote areas where grid power will never arrive. As an example, when a great drought hit the region of Sahel, in Africa, in the 1970s, father Bernard Verspieren started a program of photovoltaic water pumping to draw on water from the water-bearing stratum. In 1977, he installed the first of such devices. This is now a worldwide renowned model that became extremely popular. At that time only 10 photovoltaic water pumps were operating. Now there are tens of thousands.

Photovoltaic conversion and solar photoinduced water splitting to produce hydrogen bears a common ground in physics, and this book deals with such a scientific background. By and large, photovoltaic conversion regards seven processes (Figure 1.2), and considering particularly third-generation cells, we can name them as (0) light harvesting, (1) light absorption,(2) excited state thermalization, (3) energy diffusion, (4) charge separation, (5) charge transport, and (6) charge collection.

Each of these steps needs to be deeply understood and can be optimized, engineered, or innovated. Point (0) regards photonics, (1) regards radiation matter interaction, (2–5) concern material science and solid state physics, and (6) mainly regards interfaces. This book mainly focuses on points (1–5).

What is a book, and why in the time of Wikipedia? Of course, detailed derivations and list of notions are surpassed and obsolete. Perhaps, a simple compendium in which many different concepts and ideas are together, linked by a common purpose, is still useful. For this reason, topics here are discussed in a qualitative way, filtered by personal view. The goal is to provide a simple description, like a back-of-the-envelope description, and one can have that kind of discussion without using formal theory. Suppose a student asks you to explain a phenomenon, or

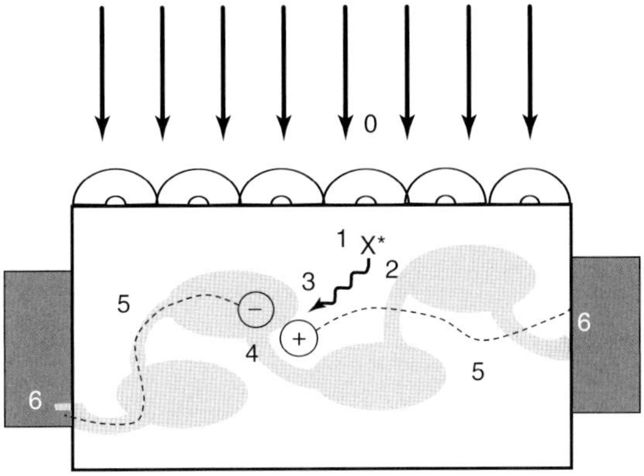

Figure 1.2 The seven processes into which photovoltaic conversion can be split, referring to a polymer bulk heterojunction cell. (0) Light harvesting, (1) light absorption, (2) excited state thermalization, (3) energy diffusion, (4) charge separation, (5) charge transport, and (6) charge collection.

you are in the laboratory and need to grasp the concept of the experiment you are doing. There is no time or space for a full theoretical derivation, yet it might be important to grasp the fundamental concept and evaluate orders of magnitude of the involved quantities. For instance, during a pump probe experiment you see triplet–triplet absorption, and knowing the triplet absorption cross section and the singlet bleaching you can estimate the formation rate and thus conjecture on the spin-flip mechanism. Reading this book should provide tools for assignment and evaluation of phenomena and an insight into the fundamental phenomena that are taking place.

2
Radiation–Matter Interaction in the Two-Level System

2.1
Introduction

Strong interaction with photons is the most relevant quality of carbon-based conjugated molecules, which determines their functions as natural chromophores. In artificial applications, the high polarizability responsible for the optical response leads to interesting electrical and nonlinear optical properties, which are exploited in photonics, optoelectronics, bioelectronics, and nanomicroelectronics. π-Electron delocalization provides many of the properties of conjugated systems. π-Electron delocalization is due to the overlap of the p-type atomic orbitals along the carbon conjugation direction. In many cases, conjugated molecules are linear, rod-shaped, or flat, with a considerable confinement of the electronic motion in two or one dimension. Correlation plays an important role in electronic dynamics, and theories are necessarily complex in order to predict correctly the electronic structure. In terms of elementary excitations, carbon-based materials are mostly excitonic, that is, support neutral states with a collective or delocalized character. The large polarizability leads to nonlinear optical properties, which manifest in a variety of phenomena, such as multiphoton absorption, Stark shift, harmonic generation, and nonlinear index modulation. Charge transfer states and polarons that come into play in solid state are major players in transport and devices, appearing as intermediate or secondary excitation under light absorption.

Here we introduce radiation–matter interaction in general terms, but mainly relating to molecular materials. Treatment is based on classical physics and macroscopic quantities, such as absorption coefficient, cross section, and so on. Then we state in simple terms and without full formalism the two-level model, to provide a microscopic picture and a physical insight.

2.2
Generality on Radiation–Matter Interaction

The electromagnetic (e.m.) field that interacts with a material can deposit energy at the characteristic frequency that defines the absorption spectrum. The electric field of the radiation, E, generates a polarization P in the material (ensemble

The Photophysics behind Photovoltaics and Photonics, First Edition. Guglielmo Lanzani.

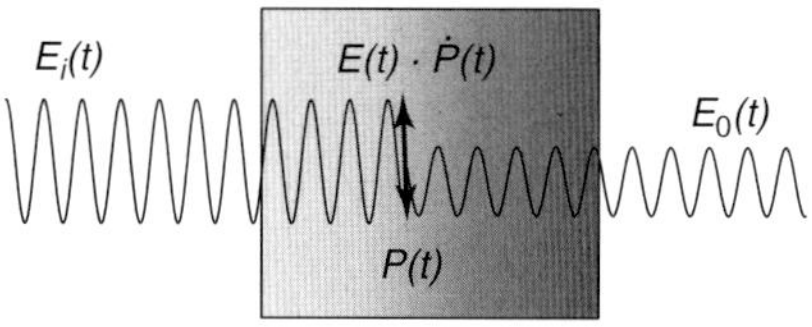

Figure 2.1 The radiation field generates a polarization $P(t)$, whose time derivative (*displacement current*) interacts with the field causing a phase shift (real refractive index) and a loss in amplitude (imaginary refractive index). The loss in radiation energy is deposited in the material. Scattering is not considered.

of molecules). The time derivative of P is a displacement current that interacts with the radiation electric field and leads to energy loss. Otherwise polarization oscillation leads to radiation re-emission with a change in phase (Figure 2.1). The average power absorbed per unit volume in the polarized medium is

$$\left\langle \frac{power}{volume} \right\rangle = \left\langle E_y(t) \frac{\mathrm{d}P_y(t)}{\mathrm{d}t} \right\rangle = \frac{1}{2} \mathrm{Re} \left[E(i\omega P)^* \right] \tag{2.1}$$

where angular bracket means time-average and time-dependent quantities are described by $\tilde{A}_y(t) = \mathrm{Re}\left[A_y(t)\mathrm{e}^{i\omega t}\right]$ with $A_y(t)$ complex amplitude for the y-component.

In general, the relationship between E and P is nonlocal in time and space. The nonlocal property in space is important only for X-ray, and it can be neglected. The nonlocal property in time should be considered, and it leads to frequency-dependent response function. Under linear approximation we write

$$P(\omega) = \varepsilon_0 \chi^{(1)}(\omega) E(\omega) \tag{2.2}$$

with the linear complex susceptibility $\chi^{(1)}(\omega) = \mathrm{Re}\,\chi^{(1)}(\omega) - i\,\mathrm{Im}\,\chi^{(1)}(\omega)$, which embodies the material properties. χ is derived by some model for the electronic structure that describes the resonance mechanism.

According to Eq. (2.2), the rate of energy loss for the e.m. field in Eq. (2.1) becomes

$$\left\langle \frac{power}{volume} \right\rangle = \frac{1}{2} \omega \varepsilon_0 \,\mathrm{Im}\,\chi^{(1)} \left| E \right|^2 \tag{2.3}$$

The intensity decay of a plane e.m. wave propagating through a medium characterized by the susceptibility $\chi^{(1)}$ is given by the Lambert–Beer law:

$$I(z) = I_0 \mathrm{e}^{-\alpha(\omega) z} \tag{2.4}$$

with absorption coefficient $-\alpha = I^{-1} \frac{\mathrm{d}I}{\mathrm{d}z}$. Energy conservation requires

$$\frac{\mathrm{d}I}{\mathrm{d}z} = -\left\langle \frac{power\ dissipated}{unit\ volume} \right\rangle = -\frac{1}{2} \omega \varepsilon_0 \,\mathrm{Im}\,\chi^{(1)} \left| E \right|^2 \tag{2.5}$$

And using $I = \frac{c\varepsilon}{2n} |E|^2$ where $\frac{\varepsilon}{\varepsilon_0} = n^2$, we find the relationship between absorption and susceptibility

$$\alpha = \frac{\omega}{cn} \mathrm{Im}\,\chi^{(1)} = \sigma(\omega) N \tag{2.6}$$

The right hand side defines absorption coefficient in a different way, often useful in spectroscopy.

$\sigma(\omega)$ is the cross section of the transition. It is expressed as units of area and represents the probability that the molecule, as a target, be hit by the incoming photon flux. The product of the photon fluence (photons per unit area and time) with the cross section gives the rate of absorption per unit molecule. N is the number of available molecules per unit volume, which is the molecular density at low excitation. At high excitation density N should be replaced with ΔN, the population difference between the ground and excited states, as it appears later.

For the sake of completeness, another observable used for measuring absorption is decadic molecular extinction coefficient, ε, defined by

$$I = I_0 10^{-\varepsilon[\mathrm{M}]z} \tag{2.7}$$

where [M] is the molar concentration (moles/liter) of the absorbing species. Since $N = N_\mathrm{A}[\mathrm{M}] \times 10^{-3}$, where N is the molecular concentration per cubic centimeter and N_A the Avogadro number ($= 6.02 \times 10^{23}$) it follows that

$$\sigma = \frac{2303\varepsilon}{N_\mathrm{A}} = 3.81 \times 10^{-19} \varepsilon\,(\mathrm{cm}^2) \tag{2.8}$$

Measurement of absorption in a solution with precise concentration can be used for experimentally evaluating the dipole moment of a molecular transition, which is an important figure of comparison with theory and also allows working out transition rates (see Strickler–Berg equation). The relationship that links the transition dipole magnitude with the integrated absorption intensity as measured in solution is

$$|\mu|^2 = 9.186 \times 10^{-3}\,\mathrm{Debye}^2\,\mathrm{mol\,l}^{-1}\,\mathrm{cm} \int \varepsilon\,\mathrm{d}\ln\overline{\nu} \tag{2.9}$$

This equation contains no local field correction, similar to those due to solvent effects. This should be kept in mind when comparing results from theoretical models that are based on parameters obtained by fitting excitation energies measured in solution, which include solvent effects. In solution, weak intermolecular interactions that occur between the solute and the solvent molecules around it (solvent cage) may shift the electronic energies. As a result, resonances in the gas phase are different from those in solution according to

$$\overline{\nu}_{0-0} = \overline{\nu}_{0-0}^{\mathrm{gas}} - K|\mu|^2 \frac{n^2 - 1}{n^2 + 2} \tag{2.10}$$

where n is the solvent refractive index, $|\mu|^2$ is the square transition dipole moment, and K is a constant inversely proportional to the molecular volume of the solute. Usually $K = 1/hca^3$, where a is the radius of the spherical cavity allocating the molecule.

The oscillator strength is a numerical quantity used to assess the intensity of a transition. For a dipole allowed transition, between state "0" and "k," here assumed along direction "x," oscillator strength is defined as

$$F_{k0} = \frac{2m\omega_{k0}}{\hbar}\left|\langle k \mid x \mid 0\rangle\right|^2 \tag{2.11}$$

This quantity specifies the absorption "area" of a transition spectral feature according to

$$F_{k0} = \frac{4.39\times10^{-9}}{n}\int \varepsilon_{k0}(\overline{\nu})\mathrm{d}\overline{\nu} = \frac{4.39\times10^{-9}}{nc}\int \varepsilon_{k0}(\nu)\mathrm{d}\nu \tag{2.12}$$

The Thomas–Reiche–Kuhn theorem states that "if Z is the number of electrons in the atomic system, the sum of the oscillator strengths on all the possible transitions is equal to Z, $\sum_k F_{k0} = Z$ if all electrons are in the ground state. For one electron $\sum_k F_{k0} = 1$, where the sum is extended onto all the possible final state "k."

The sum rule is understood also in the classical approximation. Note that radiation–matter interaction is one rare case in which quantum and classical equations are often formally equivalent. This is especially true for polarizability, dielectric function, and susceptibility. In classical terms, the sum rules regards the conservation of the number of oscillators, each quantum transition being associated with an oscillator with frequency $\omega_{0k} = \omega_j$, damping factor $\gamma = 2/T_2$ (in omega space) and amplitude weighted by $F_{k0} = f_J$. If N_j is the fraction of electrons that are resonant at $\omega_{0k} = \omega_j$, the dielectric function is

$$\hat{\varepsilon} = 1 + \frac{4\pi e^2}{m}\sum_j \frac{N_j}{(\omega_j^2 - \omega^2) - i\eta_j\omega} = 1 + \frac{4\pi e^2 N}{m}\sum_j \frac{f_j}{(\omega_j^2 - \omega^2) - i\eta_j\omega} \tag{2.13}$$

and electron conservation implies $\sum_j N_j = \sum_j N \cdot f_j = N$ or $\sum_j f_j = 1$.

The sum rule applies to the entire system, whereas a single transition can have an oscillator strength smaller or larger than one. It can also be negative, because of the frequency difference factor ω_{0k}, when downward transitions are considered (stimulated emission). The oscillator strength is a measurement of how much a transition is "allowed," because its value essentially depends on the dipole moment, which in turn depends on wavefunction overlap between the initial and final states and wavefunction symmetry. This consideration accounts for the well-known fact that charge transfer transitions, associated with displacement of an electron from one molecule or site to a neighboring site, have very little oscillator strength and huge permanent dipole moment: the wavefunction overlap is small, and the transition dipole moment is small. On the contrary, transitions between states with "similar" wavefunction extension (e.g., both localized or both delocalized on the same space volume) and proper symmetry have a large transition dipole moment and large oscillator strength.

In his textbook on optical properties of solids, Frederick Wooten introduces an example for the sum rule based on a four-level system. He arbitrarily assigns oscillator strengths for the 0–n and 1–n transitions ($n = 0, 1, 2, 3$), according to the

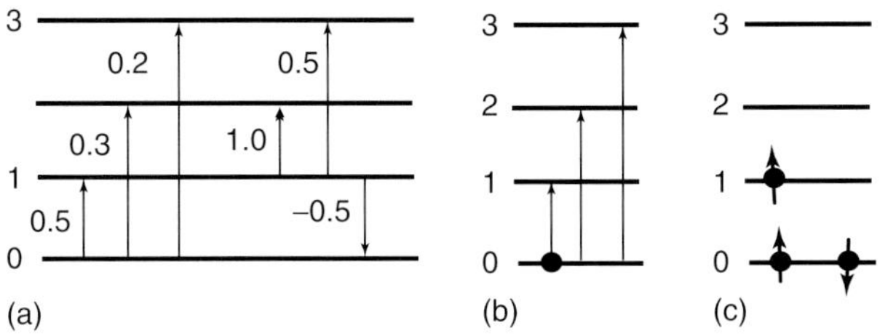

Figure 2.2 (a) Definition of oscillator strengths for transitions 0–n and 1–n and (b) only level 0 is occupied by one electron. Total oscillator strength is one. If two electrons were in the ground state, the total oscillator strength would be two. (c) Also level one is occupied, and allowed transitions are affected by Pauli exclusion principle. Total oscillator strength is three, but each electron transition sums up to different values.

rule $\sum_k F_{km} = 1$. Then he considers two examples to highlight the effect of state filling and spin rules (Pauli exclusion principle) on the sum rule. (i) The system has two electrons in level 0; the possible transitions are 0–1, 0–2, and 0–3, as shown in Figure 2.2; and the total oscillator strength is $\sum_k F_{k0} = 2$. (ii) The system has three electrons, two in level "0" and one in level "1." The partial sum gives $\sum_k F_{k1} = 1.5$ for the electron in level 1. The electron with spin up in level "0" has a total oscillator strength of 0.5, because the transition 0–1 is spin forbidden; the spin-down electron in level 0 is unaffected, and its total oscillator strength is 1. Overall, the rule $\sum_k F_{km} = 3$ is satisfied if m collects all electrons in the system.

This oscillator strength conservation rule appears in photomodulation experiments, in which the transmission changes are measured upon photoexcitation. In the transmission difference spectrum (photoexcited transmission minus dark transmission), new absorption transitions (emerging oscillator strength) in the material gap are always compensated by photobleaching (reduced oscillator strength) in the ground-state absorption. A precise balance is not expected, as in principle, the spectral range needed to fulfill the rule should extend on all allowed transitions. Hot vibronic transitions are a special case, in which redistribution of population changes the ground-state absorption (Figure 2.3, [1–4]). A molecular ensemble is in its ground state (cold). Allowed transitions are from the zero vibrational level up. Each molecule contributes to the total absorption. Owing to higher temperature or excitation, the molecular population is then redistributed onto several vibrational levels (hot state). New transitions start from higher vibrational levels, extending the absorption spectrum to the red of the original optical gap. Because the oscillator strength is constant, the total area of the differential spectrum (hot–cold) is zero.

Absorption spectra are often expressed versus wavelength; however, other units used are frequency $\nu = \frac{c}{\lambda}$, or $\omega = \frac{2\pi c}{\lambda}$, in s^{-1}, wavenumbers $\overline{\nu} = \frac{\nu}{c} = \frac{1}{\lambda}$ in cm^{-1}, energy $E = h\nu = h\frac{c}{\lambda} = \hbar\omega$ in electronvolts, where 1 eV = 8065 cm^{-1} = 1.6×10^{-19} J = 23.06 kcal/mol. Photons with energy of 1 ev have wavelength $\lambda = 1240$ nm.

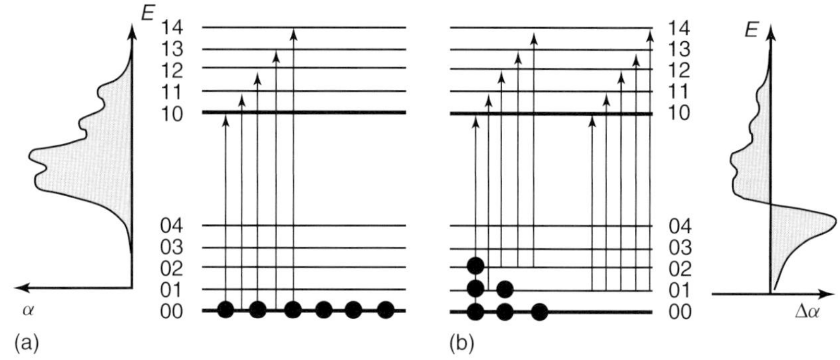

Figure 2.3 An example of vibronic molecular transitions to show the effect of constant oscillator strength. (a) Vibronic manifold of states and transitions from the lowest vibrational level. Dots here represent molecular population. The absorption spectrum is shown on the left side. (b) The hot ground state, with higher vibrational levels that are now occupied. The total oscillator strength does not change. The missing part from ground-state transition will appear in vibrational excited-state transitions. The absorption difference spectrum is shown on the right side.

2.3 Synopsis on the Two-Level Model

Electrical susceptibility is usually worked out within a model for the electronic structure. Famous and successful is the classical Lorentz model, which describes the material as a collection of damped oscillators. Each oscillator has a characteristic frequency that is associated with a feature in the optical spectrum. Better details on the process arise from the quantum mechanical description for radiation–matter interaction, which usually starts considering a two-level system (Figure 2.4). Here the energy difference $\frac{E_2-E_1}{h}$ is the frequency of the oscillator in the classical model. Each characteristic transition in the material can be described as a two-level system when coherent effects due to simultaneous excitation of several electronic transitions can be neglected.

The model system has two states: the ground state $|1\rangle$ and the excited state $|2\rangle$, at energy $E_2 - E_1 = E_0$. Wavefunctions parity is such that the dipole moment operator has a nonzero transition matrix element $\mu_{12} = \langle 2|\,\tilde{\mu}\,|1\rangle = \mu_{21}^*$ and null permanent state dipole $\langle 1|\,\tilde{\mu}\,|1\rangle = \langle 2|\,\tilde{\mu}\,|2\rangle = 0$. The system relaxation is characterized by two time constants, T_1 and T_2, called *lifetime* and *dephasing*. They have a precise physical meaning, which appears on writing the system time-evolution equation. In order to do this, we should introduce the matrix formalism and the so-called density matrix for the two-level system. It is a statistical description of the system, in the sense of the ensemble average, which finds its roots and formal definition in the theory of quantum mechanics. Some aspects of the theory are reported in Appendix 2.A, but the interested reader can find all details in virtually any book on quantum mechanics. A few examples are in the bibliography at the end of the

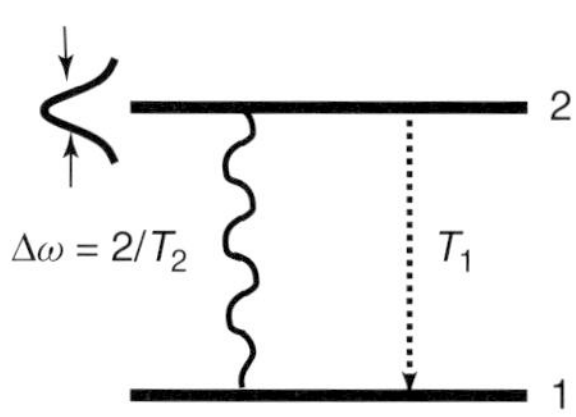

Figure 2.4 The two-level system, with schematic indication of the two relaxation processes.

chapter. For our purpose, and to quickly arrive at practical quantities, we skip the theory and report only a few definitions.

The density matrix for the two-level system is given by four time-dependent complex numbers, $\begin{pmatrix} \rho_{11} & \rho_{12} \\ \rho_{21} & \rho_{22} \end{pmatrix}$, with the following physical meaning: $\mu(t) = \mu_{21}\rho_{12} + \mu_{12}\rho_{21}$ is the time-dependent dipole moment and $P(t) = N\mu(t)$ is the total time-dependent polarization component (in a direction where μ is nonzero). N is the number of molecules per unit volume, that is, the number of two-level systems, assumed all equal, in the ensemble. N allows connecting our model to measurable ensemble quantities. $\Delta N = N(\rho_{11} - \rho_{22})$ is the population difference between the two levels. At equilibrium, the population of level "2" for the unexcited system can be assumed to be zero.

The evolution in time of the density matrix elements, as derived by the Schrödinger equation, is described by two differential equations:

$$\frac{d\rho_{21}}{dt} = -i\omega_0\rho_{21} - \frac{\rho_{21}}{T_2} + i\frac{\mu}{\hbar}(\rho_{11} - \rho_{22})E(t) \tag{2.14a}$$

$$\frac{d(\rho_{11} - \rho_{22})}{dt} = -\frac{(\rho_{11} - \rho_{22})}{T_1} + 2i\frac{\mu}{\hbar}E(t)\left(\rho_{21} - \rho_{21}^*\right) \tag{2.14b}$$

The interaction of the system with the electromagnetic field takes place through the dipole moment. Equation (2.14b) describes the generation of a population difference, according to the absorption mechanism described above, and its decay. Equation (2.14a) describes the coherent superposition of the two states, generating the oscillating polarization. Here a phenomenological decay term is included, sometimes referred to as *collisional term*. Equation (2.14) elucidates the meaning of T_1 and T_2. T_1 is the lifetime of the population difference, that is, of the excited-state population. Typically, in molecules it is a number between 0.1 and 10 ns. T_2 is the lifetime of the polarization induced in the material. Typically, in large molecules at room temperature it is 10–100 fs. Note that polarization is due to the nondiagonal terms of the density matrix, which represent the coherent superposition of the two states during the coupling with the field. T_2 is important in spectroscopy because it is responsible for the line broadening of the transition. According to this theory, the lineshape is Lorentzian. Also T_1 causes "broadening" of the resonance because of the relationship between time and energy, yet in most cases $T_1 \gg T_2$, so its contribution is negligible.

The two terms in Eq. (2.14) containing the electric field represent the excitation process, that is, how the field interacts with the system. Note that the two equations are coupled. To crack this loop one can use perturbation theory. First assuming that

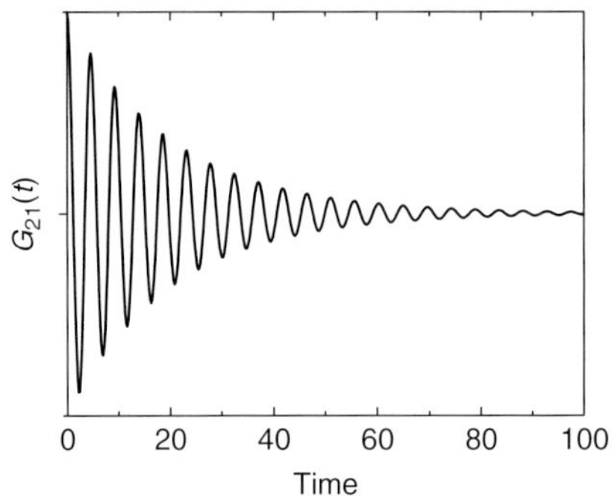

Figure 2.5 Response function for impulsive excitation of the off-diagonal term in the density matrix equation of motion.

population is known ($\rho_{11}^{(1)} = 1, \rho_{22}^{(1)} = 0$), it is possible to work out the first-order solution for polarization $\rho_{12}^{(1)}, \rho_{21}^{(1)}$. When this is used in the diagonal term equation, a second-order correction to the population $\rho_{11}^{(2)} - \rho_{22}^{(2)}$ can be obtained, and so on.

At resonance, the polarization $P(t)$ oscillates at frequency $\omega_0 = \frac{E_0}{\hbar}$, damped by T_2. The solution to the first order for the nondiagonal element, assuming that $\rho_{11} = 1$ and $\rho_{22} = 0$ is

$$\rho_{21}^{(1)}(t) = \mu_{21} G_{21}(t) \otimes E(t) \tag{2.15}$$

This is the convolution of the electric field envelope with the impulsive response function

$$G_{21} = \frac{i}{\hbar}\theta(t) \exp\left(-i\omega_0 t - t/T_2\right) \tag{2.16}$$

where $\theta(t)$ is the Heaviside function. $\rho_{12}^{(1)}(t)$ is nonresonant and will be neglected. The time dependence of $G_{21}(t)$ is shown in Figure 2.5.

The Fourier transform of Eq. (2.15), $\rho_{21}^{(1)} = \mu_{21} \frac{-1/\hbar}{(\omega-\omega_0)+i\frac{1}{T_2}} E(\omega)$, introduced into $P(t) = N\mu(t)$, considering the resonant response and using Eq. (2.2), allows obtaining the linear susceptibility of the two-level system:

$$\mathrm{Im}\,\chi^{(1)}(\omega) = \frac{\mu^2 N}{2\varepsilon_0 \hbar} g(\omega) \tag{2.17a}$$

$$\mathrm{Re}\,\chi^{(1)}(\omega) = \frac{\mu^2 N\,(\omega_0 - \omega)\,T_2}{2\varepsilon_0 \hbar} g(\omega) \tag{2.17b}$$

where $g(\omega)$ is the normalized lineshape function for "homogeneous broadening"

$$g(\omega) = \frac{\Delta\omega}{(\omega - \omega_0)^2 + \left(\frac{\Delta\omega}{2}\right)^2} \tag{2.18}$$

For a not-too-large electric field $N = N_1 + N_2 \approx N_1$. When the electric field of the exciting wave becomes large, $\Delta N \cong N_1 - N_2$ should replace N in Eq. (2.17). This happens for $\left(\frac{\mu E}{\hbar}\right)^2 > \frac{1}{T_1 T_2}$ and leads to saturation. Because ΔN is in this case a function of the field intensity, the process is, however, nonlinear in the field interaction.

The linewidth of the transition resonance is $\Delta\omega = \frac{2}{T_2}$. From this equation, it may appear easy to measure T_2, but it is not so, for it is rare that one can measure a single transition line. In most cases, the experimental absorption linewidth has inhomogeneous broadening, and by inspection one extracts only a lower limit for T_2. This is, partially, the justification for single-molecule spectroscopy.

According to the theory above, the frequency-dependent cross section of the transition is

$$\sigma(\omega) = \frac{\omega}{2c}\frac{\mu^2}{\varepsilon_0 \hbar n} g(\omega) \tag{2.19}$$

This provides a figure for the "strength" of the coupling to photons, expressed usually in square centimeters. A typical value for a large organic molecule with dipole allowed optical gap is 10^{-14}–10^{-15} cm^2.

We had stated that the polarization is decaying, but had not stated the reason. This is indeed a very complex phenomenon, hard to be modeled to the point that most of the theory includes it "phenomenologically." Any perturbation to the system (phonons and other internal degrees of freedom, collisions with other molecules from the sample and its environment, and scattering event between excited states) contributes to destroy the phase of the collective oscillating dipoles (polarization is the coherent superposition of all molecular dipoles). Those events do not change the number of molecules in the excited state but contribute to define the quality of the resonance. Highly isolated systems would have very narrow lines because of very long dephasing times. Also, the effective deactivation (population decay) of the excited state contributes to destroy the associated dipole, so there is a relationship between T_1 and T_2. Because population depends on wavefunction square (see Appendix 2.A and Eq. (2.A.7)), the relationship is

$$\frac{1}{T_2} = \frac{1}{2T_1} + \frac{1}{T_2^*} \tag{2.20}$$

T_1 is often irrelevant because it is much longer than T_2^*. However, when T_1 gets very short, population decay can contribute or even drive dephasing, $T_2 = 2T_1$. Note the apparently awkward situation here that coherence (polarization) decays two times slower than population, and it may look like one has polarization in an unexcited sample. This is, however, a mathematical paradox, for the exponential decay function never reaches zero.

What we presented above is a synopsis of the standard theory for radiation–matter interaction. Electronic transitions can be understood within this framework, which becomes more involved when dealing with real cases. Single electronic transitions are dispersed into a manifold of vibronic transitions because of vibrational normal modes (hundreds in large molecules). Inhomogeneous broadening masks the homogeneous behavior we described, and typically the lineshape is determined by the distribution of transition energy gaps (induced by disorder). Coupling with other molecules in the sample leads to the formation of extended excited states, excitons, which have a continuum of states. The formalism we presented is still generally valid, but its strict application needs some caution. For instance, each k-state of the continuum in a band can be associated with a two-level system, but the overall behavior should be understood as an integration over many transitions. Also the linear approximation may break down, as is always the case using short light pulses, calling for a higher order in the polarization versus field expansion and in the perturbation theory for the density matrix.

Except for limiting cases, the concepts of dipole moment, cross section, dephasing time, and lifetime are universally useful for a quick rationale of experiments in spectroscopy. They are part of the spectroscopy toolbox.

2.4
Absorption

Absorption can be measured using a spectrometer, which is a highly reliable user-friendly commercial equipment. The basic setup includes a light source, a monochromator, a sample holder, and a detector. There is no need to discuss this topic further, as it is a very standard one. The measured quantity is transmission (%) or absorbance. In measuring absolute values one should carefully evaluate reflectivity and scattering and eventually the substrate contribution. In general, the absorbed fraction of the incident light, I_0, at wavelength λ is

$$\Delta I(\lambda) = (1 - R(\lambda))(1 - S(\lambda))(1 - T(\lambda)) I_0 \tag{2.21}$$

where R is reflectivity, which depends on the incident angle and the index of refraction, S is the scattering contribution, and T the transmission. A careful experiment will evaluate all these contributions, measuring the specular reflectivity (i.e., that part of reflected light that follows the Snell law), the scattering fraction using an integrating sphere, and then the transmitted intensity. From the measured T, absorbance A can be obtained, and then the absorption coefficient, the absorption cross section, or the extinction coefficient. This requires some additional knowledge on sample thickness, sample density, or sample concentration. The contribution of the glass cuvette or substrate is usually subtracted during the measurement in a double-beam configuration. When this geometry is not available accurate measurements are more difficult to be obtained, simply because the light source is never stable enough to allow precise scans in series.

When transmission is not available, excitation profiles can be used to obtain a spectrum that is related to absorption, but it is not the same thing. For instance, one can measure photoluminescence (PL) at a fixed wavelength, or the integrated emission spectrum, and scan the excitation. The obtained spectrum is called *PL excitation spectrum*. It maps all the states that after excitation lead directly or indirectly to light emission. However, there can be nonemitting states that still contribute to absorption, so the excitation profile is not a copy of the absorption spectrum. This kind of measurement is used, for instance, in single-molecule spectroscopy, because emission, but not transmission, can be detected, due to experimental sensitivity. Emission has no background (to some extent and after appropriate conditions are fulfilled), whereas transmission is a change of intensity onto a large background, thus introducing an additional noise component due to background fluctuations.

An interesting experiment is PL depolarization. A linear polarized beam excites the sample, and a PL polarized beam parallel or perpendicular to it is recorded. Scanning the excitation wavelength, one collects spectra for both polarizations. Those can be a copy of one another, but can also be different. This occurs when transition dipoles are oriented in different directions for different states.

2.5 Nonlinear Absorption

At high intensity absorption becomes intensity dependent. The absorption coefficient's (α) dependence on intensity (I) is represented by the expansion

$$\alpha = \alpha_0 + \beta I + \gamma I^2 + \ldots \tag{2.22}$$

where $\alpha_0 \propto \operatorname{Im}\chi^{(1)}$, $\beta \propto \operatorname{Im}\chi^{(3)}$, and $\gamma \propto \operatorname{Im}\chi^{(5)}$, while the rate of energy loss per unit area in the material on light absorption is

$$\frac{\mathrm{d}W}{\mathrm{d}t} = AI + BI^2 + CI^3 \tag{2.23}$$

Accordingly, the nonlinear Lambert–Beer law can be easily written as

$$\frac{\mathrm{d}I}{\mathrm{d}z} = \alpha_0 I + \beta I^2 + \gamma I^3 + \ldots \tag{2.24}$$

There are many different physical mechanisms that can lead to a nonlinear response in absorption, emission, or photocurrent. Radiation–matter interaction comprises coherent multiphoton processes, such as two-photon (2PA) and three-photon transitions (3PA) that are χ^3 and χ^5 effects, respectively, which involve virtual states and two-step transitions (2SA) that involve an intermediate real state, and their effective third-order nonlinearity (in the electric field) is actually composed of the sequence of two linear, χ^1, effects. Combination of these processes is also possible, for example, a two-photon absorption (2PA) followed by a resonant one-step process (1S2PA), as described in Figure 2.6.

2PA can be detected with several experiments. Most of them require short optical pulses because 2PA depends on the intensity square. 2PA is described by the imaginary part of the third-order susceptibility. The typical term is $E \cdot \frac{\partial P^{(3)}}{\partial t} \propto \left(\operatorname{Im}\chi^3 EEE\right) E \propto I^2$. The nonlinear absorption coefficient β above is related to the imaginary part of the third-order susceptibility and, within a three-level approximation, can be written as

$$\beta = \frac{\pi\omega\left(N_{1A_g} - N_{mA_g}\right)\mu_1^2\mu_2^2}{\hbar^2\left(\Delta E - \hbar\omega\right)^2\varepsilon^2 c^2} n^2 g(\omega) \tag{2.25}$$

where a two-photon transition from $1A_g$ to mA_g takes place through a virtual state, as in Figure 2.7, and the linewidth is similar to that specified for the one-photon

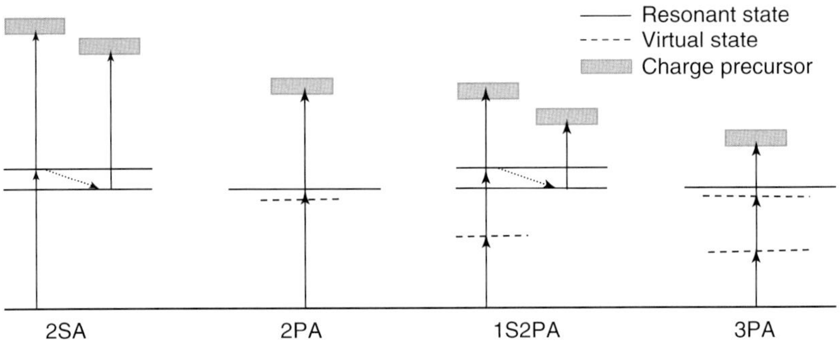

Figure 2.6 Examples of transitions leading to nonlinear absorption. 2SA = two-step absorption. 2PA = two-photon absorption. 1S2PA = one-step two-photon absorption. 3PA = three-photon absorption.

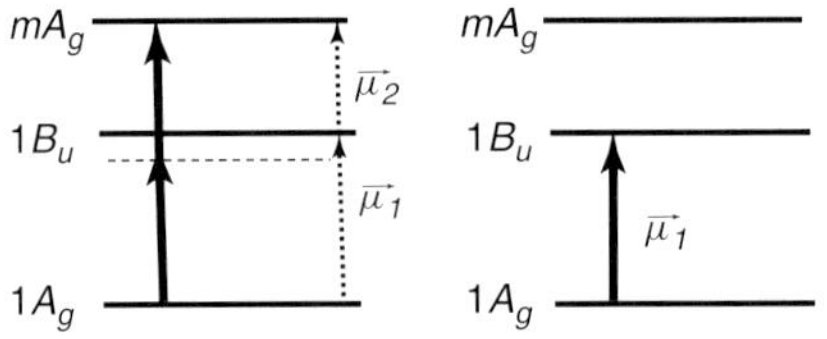

Figure 2.7 One- and two-photon transition in the three-level system with C_{2h} symmetry.

transition. Note that there is no dipole moment connecting the initial and final states.

In systems with inversion symmetry, there is an exclusion rule for one-photon allowed (ungerade or "*u*") and two-photon allowed (gerade or "*g*") states. 2PA is thus used to locate gerade states, otherwise silent in linear absorption. This is the reason why, studying a material for photovoltaic, one might like to know about its nonlinear absorption. Two-photon allowed states might play a role in energy relaxation, charge generation, or recombination. Now we review a number of experimental techniques useful for measuring nonlinear absorption. We start with single-beam techniques to measure nonlinear absorption.

1) **Nonlinear transmission** In this experiment (Figure 2.8a), transmission is measured upon changing the incoming intensity. A plot $T(I)$ will show deviation from the linear trend if 2PA, or another nonlinear mechanism, is active. This is simple to say, but difficult to measure because one need to record absolute transmission values. A careful calibration procedure and a stable laser system are needed. In addition, it is rather cumbersome to obtain a spectrum, because each spectral point will have the problem of precise normalization from one wavelength to the next.
2) **Two-photon PL excitation profile (Figure 2.8b)** This is one of the most popular techniques. Here, exciting photons with energy below half the one-photon optical gap are scanned in energy while recording PL. The excitation reaches a higher lying "*g*" state; after internal conversion, PL is detected from the lowest, "*u*," state. Interpretation of the experiment assumes that the Vavilov–Kasha

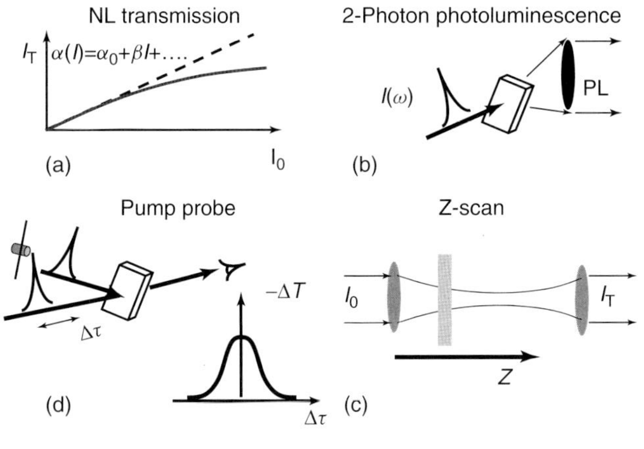

Figure 2.8 Techniques for nonlinear absorption measurements.

rule holds true. Careful intensity normalization is needed. Check of the square dependence of PL is recommended, at least for a few wavelengths, to assure that the process is indeed 2PA. Also, the PL spectrum should be fully recorded, to exclude other emission mechanisms such as phosphorescence or impurities.

3) **Electroabsorption** A cw electric field acts as perturbation on the electronic structure and can induce a shift of the energy levels (Stark or Frank–Keldish effect for discrete and continuum states, respectively) or "break" the symmetry, mixing state characters. As a consequence, "*g*" states acquire some "*u*" character, becoming partially one-photon allowed. This effect leads to the appearance of new electric-field-induced absorption bands. Electroabsorption is usually carried out with cw light. Modulation of the electric field and the lock-in technique allow detection of small changes in transmission, in the order of 10^{-4} of the linear transmission. In terms of susceptibility, this experiment is nonlinear and the signal depends on $\propto \chi^3(\omega, 0, 0)F^2$, where F is the applied field and the zeros in frequency specify the cw field. This experiment requires a device with electrodes that allow applying the electric field to the sample, possibly avoiding current injection. Usually, the signal is detected at twice the modulation frequency, according to the square dependence on the field amplitude.
4) **Photoconductivity excitation profile** Photoconductivity action spectra (i.e., the excitation profile of photoconductivity) can be used to detected "*g*" states that lead to charge carrier generation and thus current. Here again, short optical pulses are required, and the experiment is done on a device with proper electrodes for the collection of the photocurrent. It is interesting to cross check the two-photon photoconductivity experiment with that of two-photon PL. Spectra can be different, as states have different abilities to generate charges. As a very general and rough rule, one should expect currentless neutral excitation that leads to PL at low energy and more charge carrier generation going high with excitation energy, according to the autoionization picture.

Peaks, valleys, and any behavior can be due to the presence of resonances with a continuum band that favor charge separation.

A particular kind of nonlinear spectroscopy that can be envisaged here is based on recording-intensity-dependent photocurrent plots at different excitation wavelengths. Fitting the signal $PC(\omega)\sim\ \beta(\omega)I^2 + \gamma(\omega)I^3 + \ldots$ provides spectral coefficient $\beta(\omega)$ and $\gamma(\omega)$ related to two- and three-photon transitions. Plotting these coefficients versus double or triple energy, for $\beta$and γ, respectively, leads to spectra that may highlight electronic resonances associated with two-photon and three-photon allowed states.

5) **Z-scan (Figure 2.8c)** This technique is based on propagation of a Gaussian beam through the sample. In the so-called open aperture geometry it is simple and reliable. However, it requires the use of a standard reference sample of known 2PA (or nonlinear absorption) for quantitative measurements. A Gaussian beam (eventually shaped by a telescope and spatial filtered) is focused by a lens. The divergent beam from the focal point is re-collimated at a certain distance onto a detector, which sees the whole beam area. The sample is placed in the beam and its position scanned along "z," the beam propagation axis, across the focal point. Approaching $z = 0$, where the focal point is, the intensity increases upto a maximum. Accordingly, the nonlinear absorption will increase and the transmitted intensity will decrease. By recording transmission versus z-position, the detector (typically a photodiode) will measure a dip in the intensity profile at $z = 0$, due to 2PA (or any other nonlinear absorption). A 2PA spectrum can be obtained if this experiment is repeated at each wavelength and calibration is down each time. Absolute values can be obtained if a calibration sample of known 2PA cross section is used. The technique is particularly suitable for samples in solution. When using a solid sample, such as a film, a test scan at very low intensity may allow correcting the high-intensity trace for dishomogeneities of the sample, which would lead to a spurious z-dependent signal, due to unwanted linear effects.

   In the "closed" aperture scheme, when a diaphragm is placed in front of the detector, the Z-scan can also provide information on the real part of the refractive index and, in particular, on the sign of the nonlinear coefficient. This is important, for instance, to distinguish thermal effects from electronic effects. A temperature-induced change in refractive index is due to thermal expansion, leading to lower density and thus a smaller refractive index (negative nonlinear refractive index coefficient). Some years ago, the discovery of a very large nonlinear real refractive index (pointing to very large real χ^3) in Chinese green tea attracted some attention. The news was, however, trimmed down when it was demonstrated that the nonlinearity was thermal, thus with very slow response time.

   Double-beam techniques

6) **Pump probe (Figure 2.8d)** The pump probe experiment measures $\chi^{(3)}$, and thus it can detect 2PA. In pump probe (see Chapter 9 for more details), a first pulse (pump) excites the sample and a second, delayed pulse (probe) measures the pump-induced effect. A simple version is the degenerate pump

probe, where the probe is a replica of pump pulse. Transient transmission is measured for probe pulses with energy smaller than the sample optical gap. A correlation peak at $t = 0$ is the signature of 2PA (a pump and a probe photon add up to fill the gap). Two color experiments can as well be carried out, again using photons with a lower gap energy. Sometimes, when one of the two pulses, notably the pump, is absorbed linearly, two-step resonances may occur. This will appear as a peak in the $\Delta T/T$ spectrum, with an ultrashort lifetime, limited by the experimental autocorrelation. This experiment is simple, provided a good pump probe setup is available. Absolute values of the 2PA cross section require careful calibration, as in previous experiments. Pulse energy and beam area should be known precisely, especially when trying to collect a whole spectrum, and wavelength-to-wavelength normalization is required.

7) **Nonlinear photoconductivity** By using short pulses in the autocorrelation geometry, the photocurrent, generated by two time-delayed, space-coincident pulses, can be detected. This leads to second-order pulse autocorrelation (intensity correlation) mediated by charge generation. (Similarly, one can measure PL following autocorrelation excitation. Both experiments contain information on the nonlinear dynamics of the system.) In photoconductivity, the signal can never be background free. According to the equations above, and plugging in the total electric field $E_1(t) + E_2(t)e^{i\omega\tau}$ due to the light pulses delayed by τ, one finds that a 2PA signal will have a 3 : 1 peak background ratio (not taking into account chopper modulation and lock-in detection). Note that 2PA can be achieved if the semiconductor energy gap is larger than the photon energy, so in order to measure the autocorrelation of UV-visible pulses, a large band gap semiconductor is required. For a three-photon transition (3PA) the peak background ratio will be 10 : 1.

2.6 Adiabatic Approximation

The Born–Oppenheimer, or adiabatic, approximation is based on the conjecture that nuclear and electronic motions can be separated, that is, they occur in well-separated time domains. This is supported by the observation that the electronic mass is much smaller than the nuclear mass, $m_e/M_N \approx 10^{-3}$–10^{-4}, and according to classical physics the speed of nuclei will be smaller than that of electrons, $v_N \ll v_e$. Electrons move in a molecule whose nuclei are fixed.

From the quantum mechanical point of view, this translates into the separation between electronic and nuclear energies. A simple reasoning is as follows. An electron confined in a volume of radius R (molecular size) has a momentum $P \cong \frac{\hbar}{R}$ and energy $E_e = \frac{p^2}{2m_e} = \frac{\hbar^2}{2m_e R^2}$. The energy of the nuclear oscillation is $E_N = \hbar\sqrt{\frac{k}{M_N}}(n + \frac{1}{2})$. The maximum energy of the nuclear oscillation in the bound system is comparable to the electronic energy, $kR^2 = E_e$. Using $k \cong \frac{E_e}{R^2}$ one gets:

$$\frac{E^N}{E_e} = \frac{\hbar\left(\frac{E_e}{R^2 M_N}\right)^{1/2}}{E_e} = \left(\frac{m_e}{M_N}\right)^{1/2} \approx 10^{-2} \tag{2.26}$$

The emerging picture is that of well-separated electronic energies, with vibrational levels making a fine structure in between. The total energy is the sum of electronic and vibrational energies, and the total wavefunction is the product of the electronic and the nuclear wavefunctions. For a simple one-dimensional biatomic molecule (e.g., H_2^+)

$$\psi(r, R_1, R_2) = \phi(r, R_1, R_2)\chi(R_1, R_2) \tag{2.27}$$

where r is the electronic coordinate and R_k the nuclear coordinates. Generalization to the three-dimensional case and a multiatom molecule makes the expression much more involved, without important qualitative variations. The dependence of the electronic function ϕ on the nuclear coordinates is "parametric," in the sense that for each set of R's one can find a solution for $\phi_{\underline{R}}(\mathrm{r})$. The most common choice is to use the equilibrium geometry, R_{eq}, for working out the electronic spectrum.

Derivation of the molecular equation of motion under the B–O approximation is a standard topic in many textbooks, which we do not have to repeat here. We just review a few crucial points and results.

By writing the time-independent Schrödinger equation for the molecule using Eq. (2.27) for the wavefunction, one can get to a full separation of the electronic and nuclear equations of motion if a series of terms, known as *nonadiabatic terms*, are neglected. In the simple problem we are considering, those terms are

$$W = -\sum_{k=1,2} \frac{\hbar^2}{2M_k}\left(2\frac{\partial\phi}{\partial R_k}\frac{\partial\chi}{\partial R_k} + \frac{\partial^2\phi}{\partial R_k^2}\chi\right) \tag{2.28}$$

The term W in the molecular Hamiltonian is responsible for *nonadiabatic* effects, such as

1) nonradiative transitions that are induced by terms such as $\langle\phi_m|\frac{\partial^2}{\partial R_k^2}|\phi_l\rangle$ or $\langle\phi_m|\frac{\partial}{\partial R_k}|\phi_l\rangle$
2) predissociation
3) Jahn–Teller distortion
4) Herzberg–Teller (HT) transitions.

All these are accounted for by perturbation theory, mainly through the Fermi golden rule, using part of W as the perturbation operator. Those nonadiabatic effects are excluded from the calculation of the ground-state electronic structure but play a crucial role in the dynamics of the excited molecule.

According to the adiabatic approximation, each electronic state has a set of vibrational states, making up the so-called vibronic state. Excited-state energy will be characterized by the electronic quantum number, n, and a set of vibrational quantum numbers, $\{\nu_1, \nu_2, \ldots, \nu_N\}$, describing the population of each vibrational mode. A state $(n, \{v_1, v_2, \cdots v_N\})$ has energy $E_n + \left(v_1 + \frac{1}{2}\right)\hbar\omega_1 + \left(v_2 + \frac{1}{2}\right)\hbar\omega_2 + \cdots + \left(v_N + \frac{1}{2}\right)\hbar\omega_N$, where E_n is the electronic energy and $\hbar\omega$'s are vibrational energies.

2.7 Franck–Condon Principle

Under a series of approximations, as summarized below, we can proceed to evaluate the molecular absorption spectrum, introducing the Franck–Condon principle. Our strategy is to express the dipole moment matrix element. The square can then be introduced into the expression for the absorption cross section, as worked out above.

We work within the adiabatic approximation, $(m_e/M_N \ll 1 \text{ or} \Delta E_e/\Delta E_N \gg 1)$ using $\psi(r, \overline{R}) = \phi_e(r, \overline{R})\chi_v^e(\overline{R})$. Here $\overline{R}$ is a set of nuclear coordinates. The index "e" corresponds to an electronic state, while "v" is the vibrational quantum number. The dipole moment operator is

$$\hat{\mu} = -e\sum_i r_i - e\sum_k Z_k R_k = \hat{\mu}_e + \hat{\mu}_N \tag{2.29}$$

In order to keep equations simple, we assume that only one vibration has to be considered. The dipole matrix element for a transition from the electronic state g in vibrational state m to the electronic state e in vibrational state n is

$$\tilde{\mu} = \left\langle \phi_e(r, R)\chi_n^e(R)\right| \hat{\mu}_e + \hat{\mu}_N \left|\phi_g(r, R)\chi_m^g(R)\right\rangle \tag{2.30}$$

$$\begin{aligned}\tilde{\mu} = &\left\langle \chi_n^e(R)\right| \left\langle \phi_e(r, R)\right| \hat{\mu}_e \left|\phi_g(r, R)\right\rangle \left|\chi_m^g(R)\right\rangle + \\ &+\left\langle \chi_n^e(R)\right| \hat{\mu}_N \left|\chi_m^g(R)\right\rangle \left\langle \phi_g(r, R)|\phi_e(r, R)\right\rangle\end{aligned} \tag{2.31}$$

In virtue of the orthonormality rule for the electronic system, the last term on the right hand side is zero except when $e = g$. When this happens, it describes vibrational transitions, usually in the IR. We now neglect this term, keeping $e \neq g$.

$$\tilde{\mu} = \left\langle \chi_n^e(R)\right| \left\langle \phi_e(r, R)\right| \hat{\mu}_e \left|\phi_g(r, R)\right\rangle \left|\chi_m^g(R)\right\rangle = \left\langle \chi_n^e(R)\right| \mu_{eg}(R) \left|\chi_m^g(R)\right\rangle \tag{2.32}$$

where $\mu_{eg} = \left\langle \phi_e(r, R)\right| \hat{\mu}_e \left|\phi_g(r, R)\right\rangle$ is the electronic dipole moment. Owing to the dependence of the electronic wavefunction on the nuclear coordinates, the dipole moment is also coordinate dependent. This dependence is, however, weak, and the dipole moment can be expanded into a series, $\mu_{eg} = \mu_{eg}^0 + \mu_{eg}' R + \ldots$. By retaining only the first term, one adopts the so-called Condon approximation.

According to the Condon approximation, the transition dipole moment from the electronic state g in vibrational state m to the electronic state e in vibrational state n is

$$\tilde{\mu}_{\mathrm{gm,en}} = \mu_{eg}^0 \left\langle \chi_n^e(R)\right| \left.\chi_m^g(R)\right\rangle \tag{2.33}$$

The absorption cross section is proportional to the square of this quantity,

$$\sigma \propto \left|\tilde{\mu}_{\mathrm{gm,en}}\right|^2 = \left|\mu_g^0\right|^2 \left|\left\langle \chi_n^e(R)\right| \left.\chi_m^g(R)\right\rangle\right|^2 = \left|\mu_g^0\right|^2 \times FC \tag{2.34}$$

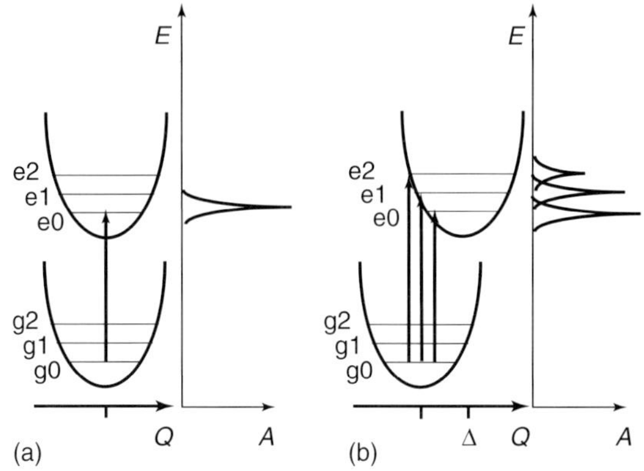

Figure 2.9 (a) Zero displacement between the ground and excited states. The molecule does not change geometry on electronic transition (zero electron–phonon coupling). Only one vibronic transition is allowed, here the 0–0, assuming only g0 is populated. "*A*" is absorbance and "*E*" energy. The expected absorption spectrum is shown. (b) Equilibrium position displacement Δ between the ground- and excited-state potentials. An arbitrary value $\Delta < 2^{1/2}$ in dimensionless unit is shown. Accordingly, the absorption spectrum shows vibrational replica. Only the first three are shown.

Here, FC stays for Frank–Condon integrals, which are overlap integrals of vibrational wavefunctions, squared.

The full expression for the vibronic transition cross section between two electronic states, ground (*g*) and excited (*e*), will be

$$\sigma_{\mathrm{A}} = \frac{\omega}{c} \frac{\mu_{\mathrm{eg}}^2 \, |\langle n|m\rangle|^2}{2\varepsilon_0} g(\omega) = \frac{\omega}{c} \frac{\mu_{\mathrm{eg}}^2}{2\pi\varepsilon_0} \frac{|\langle n|m\rangle|^2 \, \Gamma}{\left(E_{\mathrm{en}} - E_{\mathrm{gm}} - \hbar\omega\right)^2 + \Gamma^2} \tag{2.35}$$

where we simplified the expression for the FC integrals and assumed a *Lorentzian* for the lineshape function g.

Let us first consider the rare situation in which the excited state and ground state have the *same vibrational frequency* and *equilibrium geometry*. In this situation, the excited-state potential surface is a copy of the ground-state potential surface, simply vertically shifted. Owing to orthonormality of the harmonic oscillator wavefunctions, only the transitions between the same vibrational quantum numbers ($n = m$) have nonzero FC integrals. If the "0" vibrational level of the ground state is the only one populated, there is only one transition, the so-called "0–0" transition, as in Figure 2.9.

In general, however, this is not the case because the vibrational wavefunctions of the excited electronic state, *e*, are different from those of the ground state, *g*. Going from ground state to excited state many things can change, leading to nonzero FC integrals:

1) The excited-state equilibrium geometry can be different from the ground-state equilibrium geometry. The potential energy surfaces (PESs) are equal in shape but displaced along R.

2) For the same normal mode, the frequency in the excited state can be different from that of the ground state. In harmonic approximation, $\omega_e - \omega_g \neq 0$.
3) The normal mode coordinates in the excited state can be different from those in the ground state (i.e., the direction of vibrational displacement of the nuclei is changing). In this case, the new direction may be expressed by a linear combination of ground-state normal modes, an operation called *Dushinsky rotation* (Box 2.2). In most cases, however, this effect can be neglected, assuming that the shape of the molecule is unchanged. This corresponds to neglecting the Dushinsky effect.

Finally, in real-life situations the PES for molecular vibrations is anharmonic, as demonstrated by thermal expansion (see Box 2.1). Thermal expansion would not occur in a purely harmonic system, for in that case, no matter which vibrational state the system is, the equilibrium position (average internuclear distance) never changes. In spite of this "striking" evidence of ubiquitous anharmonicity, by and large the harmonic approximation is acceptable when dealing with optical transitions.

For all these reasons, FC integrals are different from zero, giving rise to the vibronic progression in the absorption spectrum. In the *displaced harmonic oscillator* model (Figure 2.9), only effect 1 is considered. Accordingly, the excited-state PES is a replica of the ground PES but displaced along the nuclear coordinate. Now vibrational wavefunctions in "g" and "e" are no more the same (because they have different equilibrium position) and, most important, do not fulfill the orthonormality rule. Overlap integrals are different from zero for any pair of vibrational quantum numbers. This develops a series of vibrational replica over which the electronic oscillator strength (which is unchanged) is distributed. Each vibronic transition has a width given by the electronic dephasing T_2 to first approximation equal for all. The amplitude of each vibrational peak is proportional to the corresponding overlap integral square, which is in turn a function of the displacement Δ. Displacement δR for the strongly active mode can be in the order of 0.01 nm. Remember here that Δ is expressed in dimensionless units $\Delta = \left(\frac{m\omega}{\hbar}\right)^{1/2} \delta R$.

In the displaced harmonic oscillator model, Franck–Condon integrals can be analytically worked out. Let us assume that only the zero vibrational level is populated in "g." The vibronic transition *0–n* has FC integral:

$$F_{n0} = \frac{e^{-S} S^n}{n!} \tag{2.36}$$

where $S = \frac{\Delta^2}{2}$ is the so-called Huang–Ryss factor. This expression obviously corresponds to the Poisson distribution, with average S. This is not incidental, and it has deep implications. Each vibrational state "n" is populated according to a weight factor corresponding to the Poisson coefficient. Having a sufficient bandwidth for excitation, all vibrational states in "e" would be populated, according to Poisson distribution. Their coherent superposition would thus be, by definition, the "coherent" state in quantum mechanics. Such a wavepacket of vibrational states would correspond to the minimum uncertainty state. This will be further

Box 2.1: Thermal Expansion

The phenomenon of thermal expansion, common in our daily life, can be accounted for only by vibrational anharmonicity. To see this, we consider a simple biatomic molecule. Introducing the anharmonic vibrational potential we show that a higher temperature corresponds to an increase in the average interatomic distance.

During vibrational stretching, the interatomic distance changes from a to $a \pm x$. The harmonic elastic force is $F = -\mathrm{K}\langle x\rangle$, and the harmonic energy associated with the oscillation is $E_{\mathrm{K}} \cong \frac{1}{2}Kx^2$. At thermal equilibrium $\langle E_{\mathrm{k}}\rangle = \langle E_{\mathrm{T}}\rangle = \frac{1}{2}K_{\mathrm{B}}T$, and the force is null. Accordingly, within harmonic approximation the average interatomic distance does not change because $\langle x\rangle = 0$. At this level of approximation thermal expansion cannot be explained.

The average quadratic displacement, however, is $\langle x^2\rangle = \frac{K_{\mathrm{B}}T}{K}$, different from zero and temperature dependent. The anharmonic elastic force is $F = -Kx + K'x^2$ with $K'/K \ll 1$. At equilibrium, the elastic force is null, $F = -K\langle x\rangle + K'\langle x^2\rangle = 0$, which implies $\langle x\rangle = \frac{K'}{K}\langle x^2\rangle$. If we use the harmonic displacement square in this expression, the average position with anharmonicity is $\langle x\rangle = \frac{K'}{K^2}K_{\mathrm{B}}T$. This result shows that whenever $K' \neq 0$ the average interatomic distance increases with temperature, that is, there is thermal expansion. This is evident in Figure 2.10, that shows the shift in equilibrium position only for the anharmonic potential.

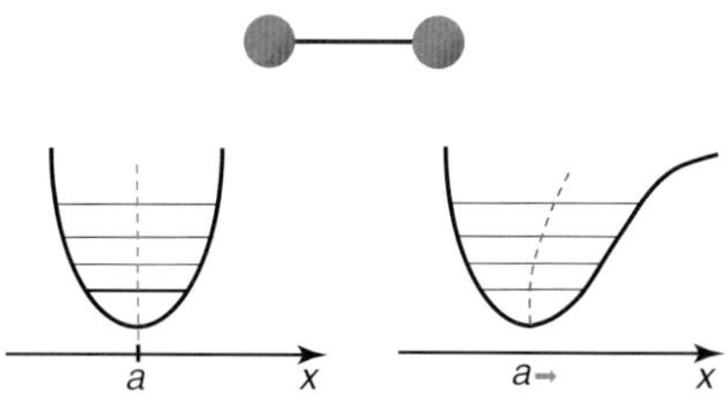

Figure 2.10 The vibrational potential for the biatomic molecule. "x" is the interatomic distance. Left: the harmonic vibrational potential. The average equilibrium position does not change with the vibrational population (temperature). Right: anharmonic vibrational potential. The average equilibrium position depends on the vibrational population. Higher lying states (high temperature) are associated with larger equilibrium distances.

mentioned when discussing the time domain radiation–matter interaction in Chapter 7.

"S" represents the average number of phonons, generated by the vibronic transition. This can be seen by recovering the expression in real space:

$$S = \frac{\Delta^2}{2} = \frac{m\omega}{2\hbar}\delta R^2 = \frac{m\omega^2}{2\hbar\omega}\delta R^2 = \frac{m\omega^2\delta R^2}{2}/\hbar\omega = \frac{\Delta E}{\hbar\omega} \tag{2.37}$$

Here ΔE is the energy shift from the bottom of the PES at displacement Δ from equilibrium. ΔE is known as *relaxation energy*, given by $\Delta E = E_{vert} - E_{00}$, where E_{vert} is the energy of the vertical transition occurring at a fixed nuclear coordinate and E_{00} the vibrationless transition. $\Delta E = S\hbar\omega$ can also be called *"reorganization" energy*, in the sense specified in Chapter 8 regarding Marcus theory. This is the energy required to distort the configuration of the ground state into the configuration of the excited state without electronic transition. If the vibrational frequency in the excited state, ω_e, is different from that in the ground state, ω_g, this term will be corrected into $\Delta E = S\left(\frac{\hbar\omega_e^2}{\hbar\omega_g}\right)$ to take into account the work needed to change frequency.

The FC integrals give the amplitude of each vibronic replica. The value of S for a single-mode vibrational progression is easily extracted from experimental spectra using $\frac{I_{01}}{I_{00}} = S$, where I_{nm} is the amplitude of the transition from vibrational level n to vibrational level m. Depending on the value of S, the vibrational progression has a different shape. For $S < 1$, the 0–0 peak is the highest. On increasing S, other peaks in the progression get higher. For a very large S, as typical for low-frequency modes, the Poisson distribution approaches a Gaussian and the highest vibrational replica is about the center of the modulated spectrum. Figure 2.11 represents vibronic absorption lineshapes for different Δ or S values.

In real-life, several modes have large enough displacement (or S factor) to give rise to sizable vibrational progression. Real spectra thus contain a congestion of

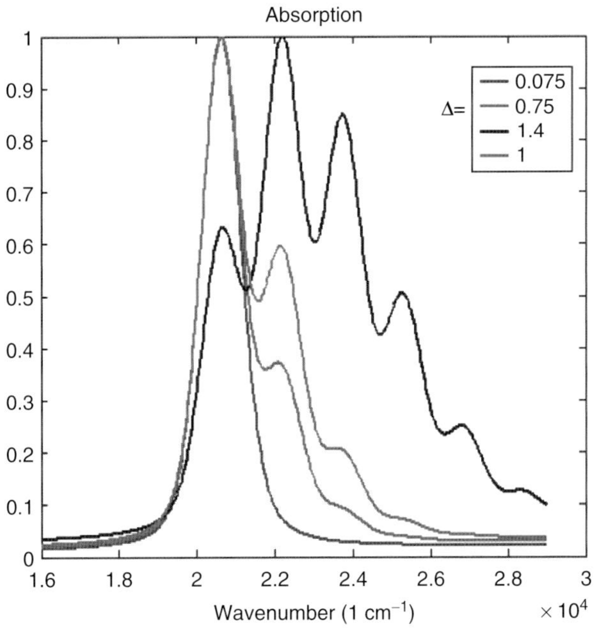

Figure 2.11 Vibrational lineshape for vibronic absorption at different displacements between ground and excited PES.

Box 2.2: Multimode Vibronic Progression

Multimode Huang-Ryss expression for two independent oscillators at ω_1 and ω_2

$$S = S_1 + S_2$$

$$I(\omega) \propto \left|\tilde{\mu}_{eg}\right|^2 e^{-S} \sum_{k_1,k_2} \frac{S_1^{k_1}\ S_2^{k_2}}{k_1!\ k_2!} \frac{\Gamma}{(\omega - \omega_0 - k_1\omega_1 - k_2\omega_2)^2 + \Gamma^2}$$

k_1	k_2	$I_{k_1k_2}$	E	$I_{k_1k_2}/I_{00}$
0	0	I_{00}	ω_0	1
0	1	I_{01}	$\omega_0 + \omega_2$	S_2
0	2	I_{02}	$\omega_0 + 2\omega_2$	$S_2^2/2$
0	1	I_{10}	$\omega_0 + \omega_1$	S_1
1	1	I_{11}	$\omega_0 + \omega_1 + \omega_2$	$S_1 S_2$

vibronic transitions, and a careful analysis is required for assignment (see example in Box 2.2). Practically, this can be done only for high-resolution spectra, obtained from isolated molecules in an inert solid matrix at very low temperatures. The analysis gives a precise determination of the difference between normal coordinate equilibrium values between ground and excited states. These normal coordinate displacements together with the vibrational frequencies give a quantitative determination of the difference between 0–0 (relaxed) and vertical (unrelaxed) excitation energies that does not require any knowledge of how the normal coordinate is described in terms of atomic displacements. The difference between the normal coordinates equilibrium values in the ground and excited states, δR, can be derived by fitting the vibronic intensity profile to obtain S and thus using Eq. (2.37).

PL spectra can be analyzed in the same way because the same theory is valid, just considering downwarding transitions from the lowest vibrational level of the excited state. PL spectra can be measured from a single molecule, and this technique is a powerful tool for obtaining well-resolved vibrational progression. PL has a vibrational structure corresponding to the modes in "g," while absorption shows the modes in "e." Comparing the vibronic modulation in absorption and emission is thus a simple way to test some of the approximations we stated above, for instance, if the vibrational frequency is the same. In general, experiments suggest that they change of few wavenumbers at most. (See the paragraph on Strickler–Berg equation for a further discussion on mirror symmetry between emission and absorption). A notable difference is seen for polyenes and aromatic molecules. In such systems the dimerization pattern (the alternance of single and double bounds) can change in the excited state. Aromatic molecules can turn into the quinoid form in the first singlet excited state, with full exchange of the single- and the double-bond positions. This leads to interesting phenomena such as large conformational readjustments (for instance, planarization) and a dramatic change of some vibrational frequencies.

2.8 Beyond Condon Approximation

To begin with, let us take one step back. FC as discussed in the previous chapter describes the effect of some particular vibration on the absorption spectrum, that is, those with an appropriate symmetry. The FC active vibrations are the total symmetric modes, or the "breathing" modes, those that do not change the shape of the molecule. These vibrations have the same symmetry of the ground (equilibrium) state and change equilibrium position on transition to the excited state. On contrary, the difference in the equilibrium position between two electronic states is null for non-total-symmetric modes. Those total symmetric, FC active modes are also shooting up in resonance Raman spectra, so they can be assigned by a careful spectroscopic investigation.

The detailed discussion on symmetry and group theory in spectroscopy is certainly out of the scope of this book; a good reference is in the bibliography. While the good spectroscopist should know this in some detail, we can attempt here a quick operative summary. Each molecule has a peculiar shape, which fulfills a set of symmetry operations (rotation, inversion, and reflection) on its nuclear coordinates (for instance, inversion is x, y, z into $-x$, $-y$, $-z$). At the end of each operation the obtained molecule should look the same as the initial one for the operation to be valid. The set of all proper symmetry operations constitutes a group, in mathematical theory. This group can be represented by a set of symmetry species, and each wavefunction, or electronic or vibrational state, can be assigned to one of those symmetry species. An important species is the total symmetric species, which is that of the molecular ground state.

We now reconsider the theory of vibronic coupling, which was developed before within the overlap view, considering the vibronic coupling operator $\frac{\partial H}{\partial Q}$ and applying perturbation theory. The potential energy for a single mode can be expanded as

$$V_e(Q) = V_e(0) + \left.\frac{\partial V_e}{\partial Q}\right|_0 Q + \frac{1}{2}\left.\frac{\partial V_e^2}{\partial Q^2}\right|_0 Q^2 \tag{2.38}$$

The linear term in Q is the force acting on the nuclei following the electronic transition. The second term is the elastic contribution and contains the force constant of the bond. In harmonic approximation higher order derivatives are null. With this interpretation we get

$$V_e(Q) = V_e(0) + \langle e| \frac{\partial H}{\partial Q} |e\rangle\bigg|_0 Q + \frac{1}{2}K_e Q^2$$
$$= V_e(0) + \frac{1}{2}K_e \left[Q + \frac{\langle e| \frac{\partial H}{\partial Q} |e\rangle\big|_0}{K_e} \right]^2 - \frac{1}{2}K_e \left[\frac{\langle e| \frac{\partial H}{\partial Q} |e\rangle\big|_0}{K_e} \right]^2 \tag{2.39}$$

Expression (2.39) is the energy of a displaced harmonic oscillator, as a perturbed state due to the vibronic coupling operator. The shift of the equilibrium position, Δ, is associated with the vibronic coupling by

$$\langle e| \frac{\partial H}{\partial Q} |e\rangle\bigg|_0 = K_e \Delta \tag{2.40}$$

So in order to have a nonzero displacement, the force expressed by the diagonal matrix element in the left term of Eq. (2.42) should be different from zero. This can occur only if Q is total symmetric, otherwise the integrated function is odd and the integral zero. Note that according to Eq. (2.39), which represents the "vibronic coupling view," the harmonic oscillator is also depressed in energy by the amount $1/2K_e\Delta^2$ with respect to the bare, zero coupling molecule.

Now we look at what happens when we consider the next term in the dipole moment expansion, $\mu_{\mathrm{eg}} = \langle \phi_{\mathrm{e}}(r, R)| \hat{\mu}_{\mathrm{e}} |\phi_{\mathrm{g}}(r, R)\rangle = \mu^0_{\mathrm{eg}} + \mu'_{\mathrm{eg}} R + \ldots$, or alternatively but exactly with the same meaning, we consider that the energy of an electronic state can change during a vibration:

$$H = H(0) + \sum_k \left(\frac{\partial H}{\partial Q_k} \right)_0 Q_k + \cdots \tag{2.41}$$

As you can see, the physics behind is still that of the vibronic coupling, and we are considering a force acting on nuclei. Now, however, we look at nondiagonal terms (before we were considering the diagonal term $\langle e| \frac{\partial H}{\partial Q} |e\rangle$). Using perturbation theory, we see that the effect of such a nondiagonal term is

1) shifting the energy of the state,
2) mixing the electronic wavefunction.

Both phenomena are caused by nuclear motion. In other terms, the molecular vibration temporarily changes the molecular shape and requires a renormalization of the electronic structure. Operatively, we first consider a frozen molecule and find its zero-order eigenstates; then we assume weak perturbation of the nuclear position and find the new first-order electronic states. The corrected wavefunctions are

$$\psi_j^{(1)} = \psi_j^{(0)} + \sum_{k \neq j} a_k \psi_k \tag{2.42}$$

where the coefficient is

$$a_k = \frac{\langle k| \sum_n (\partial H/\partial Q_n)_0 |j\rangle Q_n}{E_j - E_k} \tag{2.43}$$

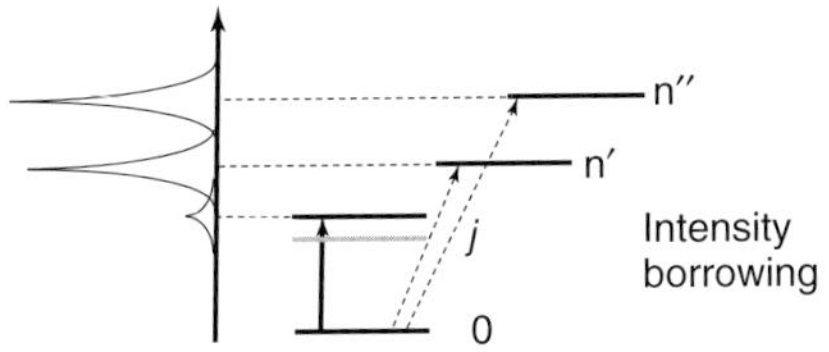

Figure 2.12 The dipole forbidden electronic transition $0-j$ acquires oscillator strength from the higher lying transitions $0-n'$ and $0-n''$ through the vibronic coupling mechanism. The absorption spectrum is shown on the left. The false origin peak is shown, while vibrational structure is neglected.

The dipole moment matrix element becomes

$$\mu_{j,0} = \langle j|\mu\,|0\rangle + \sum_n a_n^* \langle n|\,\mu\,|0\rangle = \sum_n a_n^* \langle n|\,\mu\,|0\rangle \tag{2.44}$$

Here, the first term in the right hand side is the FC contribution, as discussed above. If this term is zero, because of symmetry selection rule (dipole forbidden transition), the second term becomes important, and one sees that the transition $0{-}n$ is still possible, albeit with very weak intensity, proportional to

$$\left|\mu_{j,0}\right|^2 = \left|\sum_n a_n^* \langle n|\,\mu\,|0\rangle\right|^2 \tag{2.45}$$

This term represents the intensity borrowing mechanism due to vibronic coupling. It can appear both in absorption and emission. Essentially the transition $0-n$, dipole forbidden, "borrows" intensity from transitions to other states (n', n'' in Figure 2.12). Contribution from other states is weighted, as usual in perturbation theory, by energy denominator, so that only nearby states can give a contribution.

The vibronic coupling we described is also known as *Herzberg-Teller* coupling. It gives rise to the phenomenon of false origins. This comes from the further observation that a vibronic allowed transition (i.e., a dipole forbidden transition activated by vibronic coupling), $0{-}j$ does not occur at the E_{0-j} energy gap. In absorption, a vibrational quantum needs to be generated, so that the absorption transition energy appears at higher energy $E^{\mathrm{HT}}(0-j) > \mathrm{E}_{0-j}$. In emission, again a phonon should be generated, and the transition peak appears one vibrational quantum below the true origin of the electronic transitions. One should thus identify correctly the false origin, guess which vibrational mode is responsible for the HT activation, and retrieve the zero phonon, dipole forbidden electronic energy gap.

The vibrations that can be active in this process are *non-total symmetric*. The reason is that only those "Q" can give a nonzero off-diagonal matrix term for the vibronic coupling operator in Eq. (2.45) and the following equations. Intuitively, a non-total symmetric vibration changes the shape of the molecule, thus it can "compensate," by distorting, the asymmetry of the electronic wavefunction. It is a dynamical relaxation of the initial geometrical condition, which generates, nonpermanently, a new molecule with lower symmetry. The lower the symmetry the less strict the selection rules.

In making a spectral assignment, one should be aware that on top of a vibronically allowed, HT transition, a Franck–Condon progression due to a totally symmetric mode could develop.

Appendix 2.A: Two Level Density Matrix

A two-level system (an atom or molecule in the ensemble) in state "s" is described by wavefunction

$$\Psi_s(\bar{r}, t) = C_1^s(t)\phi_1(\bar{r}) + C_2^s(t)\phi_2(\bar{r}) \tag{2.A.1}$$

The coefficients "$C_n(t)$" describe the time evolution of the state "s," which is defined by the initial condition ($C_n^s(0)$), according to the usual quantum mechanic interpretation:

$C_1^s(t)C_1^{s*}(t)$ is the probability to be in state "1";
$C_1^s(t)C_2^{s*}(t)$ or $C_2^s(t)C_1^{s*}(t)$ is the probability that state "s" is a coherent superposition of "1" and "2";
$C_2^s(t)C_2^{s*}(t)$ is the probability to be in state "2".

The distribution of states (in a statistical sense) in the ensemble is given by a function $p(s)$. The density matrix formalism allows working out the expectation values of operators in cases where the precise wavefunction is unknown through a statistical average according to

$$\rho_{nm} = \sum_s p(s)C_m^{s*}(t)C_n^s(t) = \overline{C_m^{s*}(t)C_n^s(t)} \tag{2.A.2}$$

Even if most of the time $p(s)$ is not explicitly expressed, it is important to bear in mind the general meaning of the definition in Eq. (2.A.2): the density matrix represents the ensemble average of the system distribution.

The ensemble average of the expectation value of observable A is

$$\overline{\langle A\rangle} = \sum_s p(s)\sum_{nm} C_m^{s*}(t)C_n^s(t)A_{mn} = \sum_{nm} \rho_{nm}A_{mn} = \mathrm{tr}(\hat{\rho}\hat{A}) \tag{2.A.3}$$

where the last term on the right exploits the matrix formulation and "tr" stands for trace. The physical meaning of the density matrix elements for the two-level system is obtained according to Eq. (2.A.3).

The time evolution of the density matrix is

$$\dot{\rho}_{nm} = \sum_s \frac{\mathrm{d}p(s)}{\mathrm{d}t} C_m^{s*}(t)C_n^s(t) + \sum_s p(s)\left(C_m^{s*}(t)\frac{\mathrm{d}C_n^s(t)}{\mathrm{d}t} + \frac{\mathrm{d}C_m^{s*}(t)}{\mathrm{d}t}C_n^s(t)\right) \tag{2.A.4}$$

Assuming $p(s)$ is time independent, only the second term remains different from zero. Using the Schrödinger equation for the time evolution of the probability amplitude ($i\hbar \frac{d}{dt} C^s_m(t) = \sum_n H_{mn} C^s_n(t)$) and after some calculation (see full derivation in Ref. [5]), the general expression from which Eq. (1.7) is derived is

$$\dot{\rho}_{nm} = -\frac{i}{\hbar} \left[\hat{H}, \hat{\rho} \right]_{nm} \tag{2.A.5}$$

where the square brackets denote the usual Poisson commutator. When using $\hat{H} = \hat{H}_0 - \hat{\mu} \cdot E(t)$ in Eqs. (2.A.5) the (2.14) are obtained, after the additional inclusion of the "collision" term. The latter is done by adding phenomenological damping terms, generally defined, for the two-level system, as $\gamma_{12} = \gamma_{12} = \frac{1}{T_2}$ and $\gamma_{22} = \frac{1}{T_1}$. Here we adopt the simplification that ρ_{11} is the probabilistic occupation of the ground state and it has infinite lifetime.

The solution of Eq. (2.A.5) can be obtained by introducing the impulsive response functions, G_{22} and G_{21}

$$G_{21} = \frac{i}{\hbar} \theta(t) \exp\left(-i\omega_{21} t - t/T_2\right)$$

$$G_{22} = \frac{i}{\hbar} \theta(t) \exp\left(-t/T_1\right)$$

where $\theta(t)$ is the Heaviside function. The density operator is the convolution of the impulsive response function (the response to a δ-like excitation) with the source term, which is itself the product of the electric field and the density operator.

$$\rho_{eg}(t) = \mu_{eg} G_{eg}(t) \otimes \left\{ E(t) \left[\rho_{gg}(t) - \rho_{ee}(t) \right] \right\}$$

$$\rho_{ee}(t) = G_{ee}(t) \otimes \left\{ E(t) \left[\mu_{eg} \rho_{ge}(t) - \rho_{eg}(t) \mu_{ge} \right] \right\}$$

Both response functions are reproduced in Figure 2.A.1. These equations are mostly solved assuming that the electric field is small enough such that the perturbation theory can be applied, $\rho(t) = \rho^{(0)}(t) + \rho^{(1)}(t) + \rho^{(2)}(t) + \ldots$. This allows working out the solution of order $p+1$ from the solution of order p through an iterative approach. The perturbation expansion leads to an alternating development crossing polarization (coherence) and population terms, such as $\rho^{(0)}_{11}(t) \rightarrow \rho^{(1)}_{21}(t) \rightarrow \rho^{(2)}_{22}(t) \rightarrow \rho^{(3)}_{21}(t) \rightarrow \cdots$.

We now derive Eq. (2.20). We write the time evolution of the state population as

$$\left| C_2(t) \right|^2 = \left| C_2(0) \right|^2 e^{-t/T_1} \tag{2.A.6}$$

from which the amplitude dependence is $C_2(t) = C_2(0) e^{-i\omega_2 t} e^{-t/2T_1}$ and similarly $C_1(t) = C_1(0) e^{-i\omega_1 t}$ where we explicitly write the phase of the time-dependent coefficient, with $E_2 = \hbar\omega_2$. The coherence term is then

$$C_1(t) C^*_2(t) = C_1(0) C^*_2(0) e^{-i(\omega_1 - \omega_2)} e^{-t/2T_1} \propto e^{-t/2T_1} \tag{2.A.7}$$

Equation (2.A.7) shows that coherence decay at half the rate of population, when population decay is the only source of dephasing.

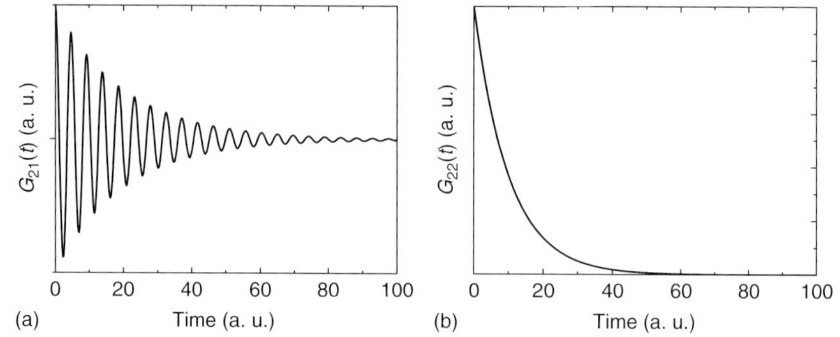

Figure 2.A.1 The impulsive response functions for the two-level system, on arbitrary scales. We assume $T_2 = 2T_1$.

Appendix 2.B: Perturbation Theory Quick Guide

The first problem with a zero-order solution for energy and wavefunctions is $\widehat{H_0}\,|n\rangle = E_n^0\,|n\rangle$. A perturbation H' is acting on the system, so the total energy operator is $\widehat{H} = \widehat{H}_0 + \widehat{H'}$. The first-order correction to energy is $E'_n = E_n^0 + H'_{nn}$, where the diagonal perturbation matrix elements are $H'_{nm} = \langle n|\,\widehat{H}'\,|m\rangle$. If diagonal terms are zero, there is no correction to first order in energy.

The first-order correction to wavefunctions is $\left|n'\right\rangle = \left|n^0\right\rangle - \sum_{k=n} a_k \left|n_k^0\right\rangle$, where the expansion coefficient is $a_k = \frac{H'_{nn}}{E_n^0 - E_k^0}$.

The Fermi golden rule expresses the transition rate in the presence of sinusoidal time-dependent perturbation as

$$W_{nk} = \frac{2\pi}{\hbar}\left|\hat{H}'_{nk}\right|^2 \rho(E_{nk})$$

where ρ is the density of the final states.

References

1. Garbugli, M., Virgili, T., Schrader, S., and Lanzani, G. (2009) *J. Mater. Chem.*, **19**, 7551–7560.
2. Kasha, M. (1950) *Discuss. Faraday Soc.*, **9**, 14.
3. Beer, M. and Longuet-Higgins, H.C. (1955) *J. Chem. Phys.*, **23**, 1390.
4. Englman, R. and Jortner, J. (1970) *Mol. Phys.*, **18**, 145.
5. Robert W., Boyd. Non linear optics Academic Press Inc.

3 Molecular Exciton

3.1 The Molecular Exciton in Aggregates

So far we considered absorption by a single molecule. Now we look at what happens when molecules are in contact with each other in an aggregate or solid. In general, even very weak interactions that occur before wavefunction overlap and covalent bonding take place can have dramatic effects on the absorption spectrum. This is because dipole moment in allowed transitions can be strong, and their peculiar long-range interaction can bring about Coulomb coupling effects.

The simplest system to study, yet very instructive, is the physical dimer. A physical dimer is made up of two molecules bound by weak dipole–dipole (van der Waals) forces, without a chemical bound. In this situation, we can discuss the result of the exciton model, which is an example of state interaction theory. At first we will skip the formal quantum mechanics, which is based on the perturbation theory. Exciton coupling is a typical phenomenon that can be grasped on simple reasoning, to obtain qualitative information that can help in understanding experimental results. Here we will follow the point dipole interaction model developed by Michael Kasha. As usual, it stays true that precise quantitative prediction can be obtained only by refined quantum mechanical simulations.

First, let us take into account that having a second molecule near a first molecule affects the overall dielectric environment, leading to a shift in the energy states. The shift is to the red, featuring a weak attractive interaction in both the ground state (D^0) and the excited state (D^*). Figure 3.1 shows the combined effect into a unique shift, D, of the energy gap of the dimer with respect to the monomer. Because there are two molecules in the dimer, the absorption intensity is doubled with respect to the monomer, and at this level of approximation the dimer state is doubly degenerate.

Now let us consider that absorption of light occurs in the dimer. In the presence of a resonant electromagnetic field, both monomers composing the dimer will develop a transition dipole moment. Such dipoles oscillate one near the other in the dimer and can interact. To describe this interaction we adopt the point dipole approximation. If they are in phase, as in Figure 3.2a, the dipole–dipole interaction is repulsive. This pushes up the energy of the dimer state, adding a positive term. On the contrary, when the two transition dipoles are out of phase, their interaction

The Photophysics behind Photovoltaics and Photonics, First Edition. Guglielmo Lanzani.

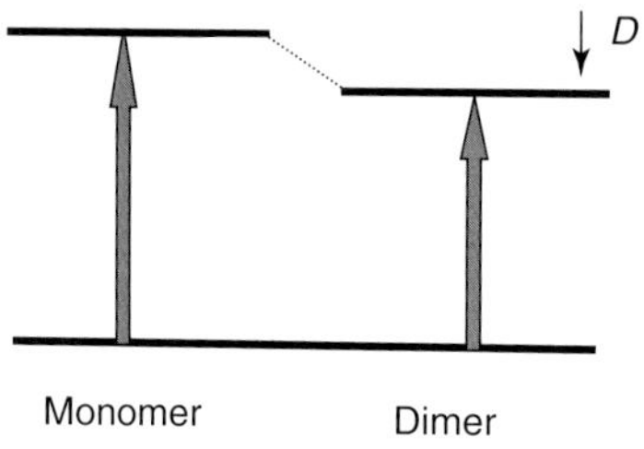

Figure 3.1 Going from monomer to dimer the electronic transition energy is reduced by a rigid shift in the state energy, *D*, here exaggerated. In the dimer, at this approximation, the electronic state is doubly degenerate.

is attractive and there is lowering in dimer state energy, as seen in Figure 3.2b. The resonance interaction thus leads to a splitting that removes degeneracy. In the dimer there are two new "collective" excited states, shared by both molecules. Localization of the energy on one or the other molecule (which are identical) is lost, and the amplitude of the two new excited states is equally distributed among the two molecules.

As long as the point dipole approximation holds true, a simple electrostatic interaction model may provide quantitative information about (i) the magnitude of the resonance splitting, (ii) the oscillator strength of the dimer transitions, and (iii) the polarization of the dimer transitions. The point dipole approximation is valid as long as $R \gg \mu/e$, where R is the intermolecular distance (respect to molecular center), μ the transition dipole moment, and e the electronic charge. In this theory, knowing the molecular transition dipole moments is a prerequisite. The excited-state resonance interaction, V, is approximated by the electrostatic interaction between dipole moments:

$$V = \frac{(\overline{\mu}_1 \cdot \overline{\mu}_2)\, |r_{12}|^2 - 3(\overline{\mu}_1 \cdot \overline{r}_{12})(\overline{\mu}_2 \cdot \overline{r}_{12})}{|r_{12}|^5} \tag{3.1}$$

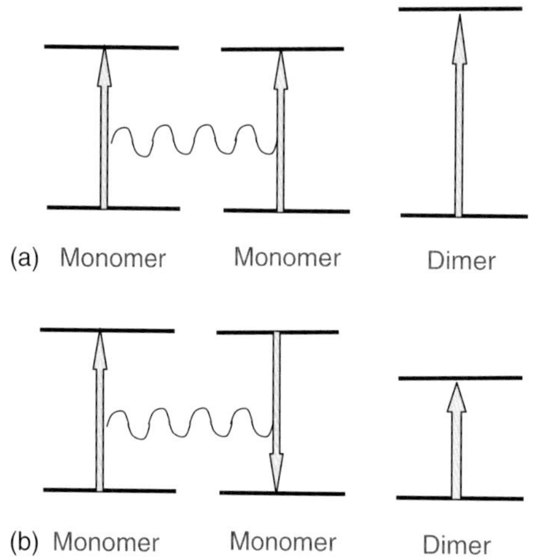

Figure 3.2 Transition dipole interaction leads to splitting of the optical gap.

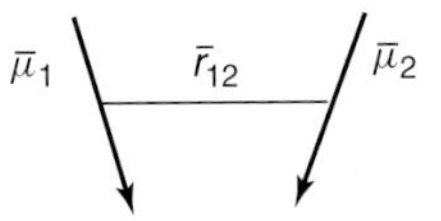

Figure 3.3 The two transition dipoles in vector representation with specified interaction.

where $\mu_{1,2}$ is the transition dipole moment for molecule 1 and 2, and r_{12} is the vector distance between the center of the molecules (Figure 3.3). For an allowed transition with oscillator strength about 1, the estimated square dipole moment is

$$\mu^2 = \frac{3e^2\hbar}{4\pi m_e \nu} = 6.7 \times 10^{-58}\ (\text{cm})^2 \tag{3.2}$$

using $\nu = 10^{15}\ \text{s}^{-1}$. For a distance $R = 0.5$ nm, the interaction energy is $V = \frac{1}{4\pi\varepsilon_0}\frac{\mu^2}{R^3} \cong 0.3$ eV. This is a reasonable order of magnitude for the resonance splitting.

The dipole moment of the dimer transitions is the vector sum of the molecular dipole moments:

$$\overline{\mu}_\text{D} = \frac{1}{\sqrt{2}}(\overline{\mu}_1 \pm \overline{\mu}_2) \tag{3.3}$$

where + and – refer to the two split-off dimer states. We discuss now a few specific molecular arrangements in space.

3.1.1 Parallel Dipoles (H-Aggregate)

This case is depicted in Figure 3.4.

The interaction energy, assuming in-phase dipoles, is $V = \frac{\overline{\mu}_1^2}{|r_{12}|^3} > 0$, repulsive.

The two dimer states, for in-phase [+] or out-of-phase [−] dipoles, have energy:

$$E'' = (E_1^* + E_2) + D + V \quad [+] \tag{3.4a}$$

$$E' = (E_1^* + E_2) + D - V \quad [-] \tag{3.4b}$$

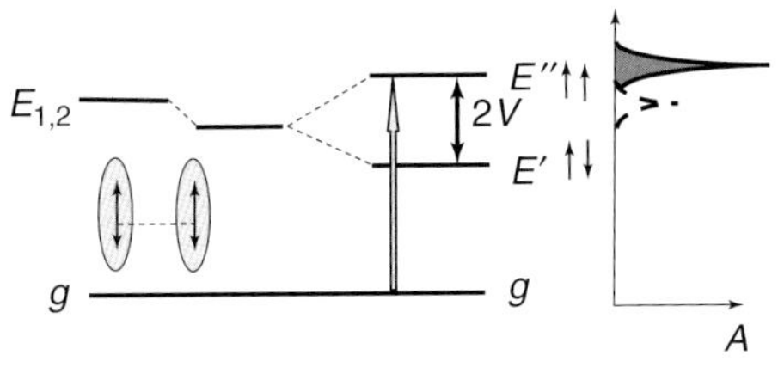

Figure 3.4 The H-dimer. Light gray shapes show the molecular arrangement. Second-order (van der Waals) effect shifts the monomer state preserving degeneracy. Exciton interaction splits the monomer transition, lifting degeneracy. The small arrows represent dipole orientation. The gray area peak is the new dimer absorption; dashed line represents the original monomer transition.

with $E'' > E'$. The two transitions have dipole moments:

$$\overline{\mu}''_D = \frac{1}{\sqrt{2}}(\overline{\mu}_1 + \overline{\mu}_2) = \sqrt{2}\overline{\mu}_1 \quad (3.5a)$$

$$\overline{\mu}'_D = \frac{1}{\sqrt{2}}(\overline{\mu}_1 - \overline{\mu}_2) = 0 \quad (3.5b)$$

Therefore, the model can also predict the strength of the transitions: the higher lying state is dipole allowed, while the lower lying state is dipole forbidden (dark). The total absorption intensity is proportional to the square of the dipole moment, $\overline{\mu}_D^2 = 2\overline{\mu}_1^2$ and is exactly twice the intensity of the monomer transition. Aggregation has lead to the interesting phenomenon of oscillator strength concentration. In addition, according to the Kasha rule, the higher lying state is essentially non-emitting because of the fast relaxation to the lower state, while the lower state is dipole forbidden, so no emission is expected from the H-aggregate configuration. In practice, some very weak emission can be detected from E′, because of misalignment of the dipoles in the aggregate or because of vibronic activation. From this we learn that managing intermolecular arrangement allows tuning energy resonances and optical properties, thus offering a handle to molecular-based technology.

3.1.2
In-Line Dipoles (J-Aggregate)

The interaction energy for in-phase dipoles is $V = -2\frac{\overline{\mu}_1^2}{|r_{12}|^3} < 0$, attractive (Figure 3.5). The two dimer states have energy

$$E' = (E_1^* + E_2) + D + V \quad [+] \quad (3.6a)$$

$$E'' = (E_1^* + E_2) + D - V \quad [-] \quad (3.6b)$$

and the two transitions have dipole moments (and thus oscillator strengths):

$$\overline{\mu}'_D = \frac{1}{\sqrt{2}}(\overline{\mu}_1 + \overline{\mu}_2) = \sqrt{2}\overline{\mu}_1 \quad (3.7a)$$

$$\overline{\mu}''_D = \frac{1}{\sqrt{2}}(\overline{\mu}_1 - \overline{\mu}_2) = 0 \quad (3.7b)$$

Here the situation is reversed: the lower state, attractive interaction, is the [+] state with non-null dipole, while the higher state, repulsive, is dark. So the model predicts redshift of the absorption and redshifted photoluminescence (PL). Because

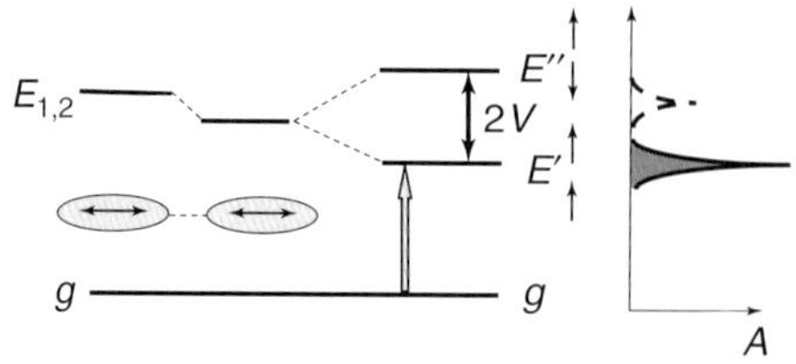

Figure 3.5 The J-dimer.

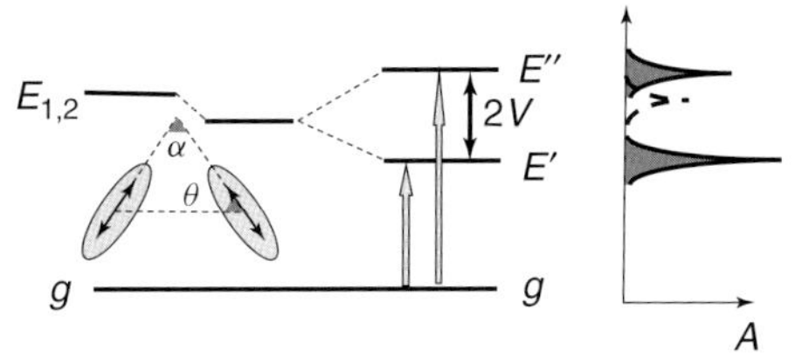

Figure 3.6 Oblique dipoles.

the radiative time is proportional to the square of the dipole moment, it turns out that the radiative time of the dimer transition is half of that of the monomer transition.

Now we are experts, and we can go faster with other arrangements.

3.1.3 Oblique Dipoles

$$V = \frac{\overline{\mu}_1^2}{|r_{12}|^3}\left(\cos\alpha + 3\cos^2\theta\right) \tag{3.8}$$

For the "+" state $\overline{\mu}_D^+ = \sqrt{2}\overline{\mu}_1 sen\theta$, for the "−" state $\overline{\mu}_D^- = \sqrt{2}\overline{\mu}_1\cos\theta$. Note that we get back H-aggregate for $\alpha = 0, \theta = \pi/2$, and J-aggregate for $\alpha = \pi, \theta = 0$. According to the vector sum rule, the polarization of the two transitions is orthogonal (Figure 3.6).

3.1.4 Coplanar Dipoles

$$V = \frac{\overline{\mu}_1^2}{|r_{12}|^3}\left(1 - 3\cos^2\theta\right) \tag{3.9}$$

This configuration (Figure 3.7) highlights the role of mutual position in controlling the interaction energy. For $\theta = 54.7^\circ$, $V = 0$, and there is no interaction in spite of the closeness of the molecules (within dipole approximation). One transition is allowed ($\overline{\mu}_D = \sqrt{2}\overline{\mu}_1$), the other forbidden. Again we go back to previous cases for $\theta = 0$, J-dimer and $\theta = 90^\circ$, H-dimer.

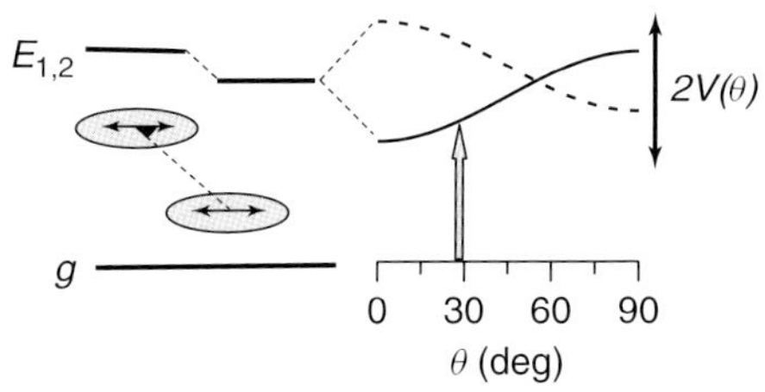

Figure 3.7 Coplanar dipoles.

3.1.5 3D Dipole Geometry ($\theta = 90°$)

$$V = \frac{\overline{\mu}_1^2}{|r_{12}|^3}(\cos\alpha) \tag{3.10}$$

$$\overline{\mu}'_D = \sqrt{2}\overline{\mu}_1 sen\frac{\alpha}{2} \tag{3.11a}$$

$$\overline{\mu}''_D = \sqrt{2}\overline{\mu}_1 \cos\frac{\alpha}{2} \tag{3.11b}$$

For $\alpha = 90^\circ$, the interaction is null, and the two transitions degenerate; for $\alpha = 0$, we get the H-dimer with E$''$ fully allowed and E$'$ forbidden (Figure 3.8).

Concluding, the molecular exciton theory of Kasha, in the point dipole approximation, allows to picture in simple terms the effect of weak intermolecular interactions. In soft matter those interactions play the major role, and many natural systems, such as the light harvesting compounds in bacteria and plants or the photosynthetic reaction centers, exploit those phenomena for tuning resonance interaction between subunits and controlling "vectorial" energy transfer. In organic semiconductors also these interactions lead to solid-state effects such as spectral shifts, energy migration, and emission quenching.

When the intermolecular distance, R, gets shorter than the dipole length, $R < \mu/e$, the dipole approximation breaks down. The full Coulomb interaction, expanded at higher order, could provide a better description, while quantum mechanics on a proper basis set is needed for a quantitative estimate of the coupling. The flavor of the involved physics is, however, well grasped by the simple point dipole model we reported.

Quantum mechanics provides a more general description of the phenomenon, preserving a quite strict correspondence with the point dipole interaction model. In the approximation of negligible ground-state interaction, a good guess for the ground-state wavefunction of the dimer is the product function$\Psi = \phi_1\phi_2$. When considering the first excited (one-exciton) state, either molecule 1 or molecule 2 could be in the excited state, with the other molecule in the ground state. If the total Hamiltonian, H, contains an interaction term V, $H = H_1 + H_2 + V_{12}$, the simple

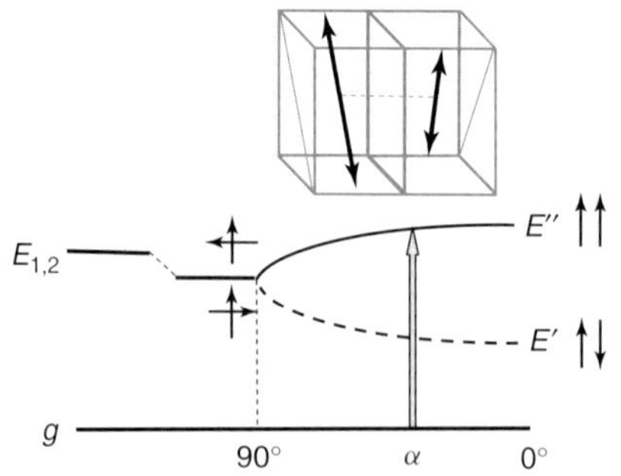

Figure 3.8 3D dipole geometry ($q = 90^\circ$).

product basis set cannot diagonalize the energy matrix.

$$E = \begin{pmatrix} H_{11} & H_{12} \\ H_{21} & H_{22} \end{pmatrix} \quad (3.12)$$

and

$$H_{11} = \langle \phi_1 | H_1 | \phi_1 \rangle \quad (3.13a)$$

$$H_{22} = \langle \phi_2 | H_2 | \phi_2 \rangle \quad (3.13b)$$

$$H_{12} = \langle \phi_1 | H_{12} | \phi_2 \rangle = H_{21}^* \quad (3.13c)$$

are all different from zero. Diagonalization leads to new states and respective energies, according to

$$\Psi_e' = \frac{1}{\sqrt{2}} \left(\phi_1^* \phi_2 + \phi_1 \phi_2^* \right) \quad E_e' = H_{11} + H_{12} \quad (3.14a)$$

$$\Psi_e'' = \frac{1}{\sqrt{2}} \left(\phi_1^* \phi_2 - \phi_1 \phi_2^* \right) \quad E_e'' = H_{11} - H_{12} \quad (3.14b)$$

The new states so obtained are the molecular exciton basis set and represent coherent superposition of singly excited local states. Each of them is a stationary solution associated with a fully delocalized state involving both molecules. The [+] and [−] states of the dipole model above are now the symmetric and antisymmetric wavefunction composition, with the node in the wavefunction representing the interdipole phase.

The simultaneous excitation of both exciton states, if selection rules allow this, would result in a coherent superposition of excitonic states, with total wavefunction $\Psi_S = c'\Psi_e' e^{-i\frac{E_e'}{\hbar}t} + c''\Psi_e'' e^{-i\frac{E_e''}{\hbar}t}$. The amplitude square $\Psi_S \Psi_S^*$ of the coherent state has stationary diagonal population terms and oscillatory nondiagonal terms that represent quantum dynamics, at characteristic frequency $|E' - E''|/h$ (Box 3.1). This coherent state regards the oscillation of the excitation probability back and forth between the two monomers. Such coherence can be generated by a short enough excitation in time. In the absence of any "bath," that is, an environment of weakly interacting subsystems such as solvent molecules, dephasing does not occur and the system stays in the coherent superposition. In reality, it would decay at a rate determined by the interaction with the bath (Box 3.1). We have now introduced two types of coherence. One is the coherent superposition of the local state to form the exciton, the other is the coherent superposition of exciton states leading to periodic phase oscillation in the total wavefunction. Both processes, when associated with a dissipative mechanism, lead to energy transfer. In general, quantum coherent transport is much more efficient than the incoherent energy hopping as described below according to Foerster theory. To explain better the coherent energy transfer process, consider the following two situations. First, the exciton state wavefunction can suddenly collapse and become localized because of scattering with some perturbation or interaction with the bath. If the perturbation is internal to the system (phonons), this process is called *self-trapping*. As a consequence, the energy, initially delocalized, is now present in one of the two molecules. This represents a displacement of energy with respect to the initial state.

Box 3.1: Excitonic Coherence

We consider a simple dimer with excitonic states Eq. (3.14) $\Psi_1 = \frac{1}{\sqrt{2}}\left(\phi_1^*\phi_2 + \phi_1\phi_2^*\right)$ with energy $E_1 = H_{11} + H_{12}$ and $\Psi_2 = \frac{1}{\sqrt{2}}\left(\phi_1^*\phi_2 - \phi_1\phi_2^*\right)$ with energy $E_2 = H_{11} - H_{12}$. The quantum superposition of the two excitonic states has a time evolution according to $\Psi_S(t) = c_1\Psi_1 e^{-i\frac{E_1}{\hbar}t} + c_2\Psi_2 e^{-i\frac{E_2}{\hbar}t}$. The probability square of such state contains stationary and oscillatory terms, easily represented by the density matrix for the system:

$$\rho = \begin{pmatrix} |c_1| & c_1^* c_2 e^{-i|E_1-E_2|/h} \\ c_2^* c_1 e^{+i|E_1-E_2|/h} & |c_2| \end{pmatrix}$$

The diagonal density matrix elements, that is, populations, are stationary against the coherent excitonic Hamiltonian dynamics, while the off-diagonal density matrix elements have a time-dependent phase that embodies the quantum dynamics. As a result, coherent wavelike dynamics depend on the existence of the excitonic coherence.

A conservative view predicts that such excitonic coherence quickly decays in disorder systems or systems embedded in complex environments. A notable example is that of pigment molecules in photosynthetic complexes. The pigment molecules are arranged in space according to an ordered geometry and are closely packed. Their interaction will thus lead to excitonic states. Accordingly, the diagonal states of the excitonic density matrix have been included in the theory once the exact location of the pigment molecules has become available from X-ray diffraction data of crystal complexes. Only recently, however, advanced time resolved spectroscopy has pointed out the role of coherences, that is, of the off-diagonal states. Against the conservative view, which predicted a quick dephasing of such coherences, it appears that phase relation is preserved for a long time. The nature of such coherences is under investigation. Whether this has a functional role in the photosynthetic complex or is just an accident has yet to be understood.

This discussion is extracted from recent works published by Fleming *et al.*, as reported in the bibliography section.

The second case is the coherent superposition, when wavefunction amplitude oscillates back and forth. Owing to dephasing and dissipation (energy sink) the probability amplitude can at some time stop the oscillatory motion and dwell on one of the two molecules. Again, this is energy relocation. This transfer mechanism may sound trivial in this example, but in larger aggregates or crystals it becomes crucial. Look at the example in Figure 3.9. Energy from the antenna system A is transferred to B1. However, B1 is not isolated, but coherently coupled to the other molecules in the aggregate. Energy from A is thus transferred to an exciton state, that is, a coherent superposition of local wavefunctions in B (the bridge system). Since there is a strong coherence between these five sites, some final state F coupled with any of the five sites will be effectively mixed with the whole set of exciton

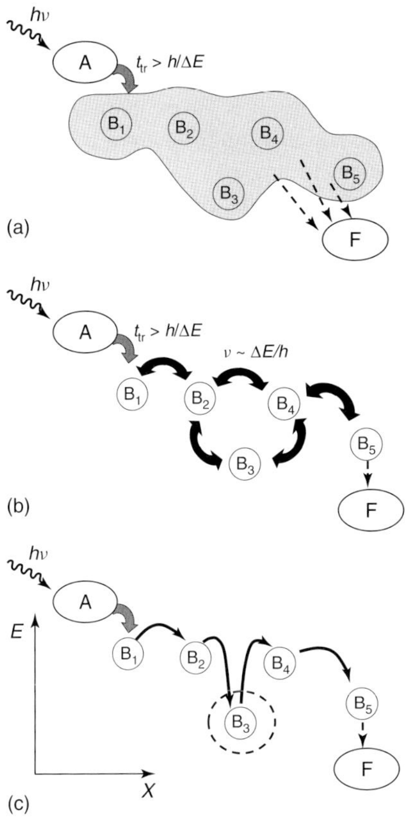

Figure 3.9 (a) Energy transfer from A to F is via the coherent superposition of B(*n*) states in an exciton. Each site is in a different position in 1D space, and energy levels are encoded in the vertical height. B_3 has the lower energy in the aggregate. The shaded area represents the lower energy exciton state wavefunction. (b) As in A, but here the initial state is a coherent superposition of exciton states (wavepacket), which oscillate among the B aggregates. (c) Energy transfer is by hopping between uncorrelated sites from A to F via B(*n*). The lower energy site B_3 may become a trap, halting the process.

states (the amount of mixing depending on the size of the energy gap between the excitonic state and the F state). In other words, some probability amplitude is *instantaneously* distributed to all B_n molecules in the aggregate. If one of these molecules is more strongly coupled to the final state, for example, molecule B_5, this will affect energy transfer to F. What does it mean *instantaneously*? We should be more precise here. If the transfer A → B occurs on a long time, compared to the resonance transfer time in the coherent state ($t_{tr} > h/\Delta E$ where ΔE is the exciton splitting) then a single exciton state is populated at the end of the transfer. If the energy is transferred from A to B in a very short time $t_{tr} < h/\Delta E$ then a coherent superposition of excitonic states is formed, that is, a wavepacket. This will coherently oscillate back and forth among the molecules in the aggregate, with frequency $\Delta E/h$. In both cases, within a time $h/\Delta E$ energy reaches B_5 in very efficient way. If B_5 is coupled to C, energy transfer may occur through the coherent state. Note, however, that in this picture energy transfer may occur from any site in the bridge system, because of the coherent mixing. It is the exciton state that transfers the energy in a much more efficient way then an incoherent transport through the bridge sites B_n by intersite hopping. In the latter case trapping may occur, for instance, in B_3, which may act as an energy sink, stopping the transfer (Figure 3.9c). Decoherence is ruling the process, dictating the lifetime of the coherent state, together with dissipative processes. The former affects only the phase among the states; the latter leads to energy redistribution. The final delivery of energy to the localized state F requires both. For many years it was believed that in complex systems such as biological compounds or disordered solids, coherence could not survive for more than a few femtoseconds. As a consequence, coherent processes were disregarded. However, recent experiments reveal that this is not the case, pointing to the role of quantum superposition of states in energy flow. Among others, the example of light harvesting systems LH-II and LH-I in photosynthetic bacteria is particularly instructive (See Box 3.2).

Box 3.2: Quantum Effects in Biology

The 22nd Solvay Conference on Chemistry "Quantum Effects in Chemistry and Biology", held in Brussels (B) in 13–16 October 2010, focused on quantum effects in biology. The topic was introduced by the chairman, Graham R. Fleming. In his address Fleming pointed out how the view on the importance of quantum effects in biology has changed with time. The old view was that quantum mechanics is important for isolated systems in pure states but can be disregarded in large systems in statistical states at thermal equilibrium. The new, emerging view is that phase coherence (Box 3.1) and quantum effects stay crucial also in large biological complexes. There are several phenomena that can be reconsidered under this view: light harvesting in photosynthesis, vision, electron- and proton-tunneling, olfactory sensing, and magnetoreception. It is not surprising that closely packed conjugated molecules have excitonic interaction in aggregates. Accordingly, stationary states (diagonal in the density matrix formalism) have been included in the theory of all the aforementioned phenomena. The new result is that ultrafast nonlinear spectroscopic experiments (mainly the so-called 2D spectroscopy) point out the persistence of coherent superposition of quantum states also in large biological complexes. Phenomena such as energy transfer (light harvesting), electron or proton transfer (photosynthesis), isomerization (early vision event), long-range charge transport (photosynthesis), or radical pair evolution (magnetoreception) could in principle be affected by coherence. Light harvesting was discussed in Box 3.1. Electron tunneling regards the coherent displacement from one location to another through a nonclassical state (typically through a barrier) as an alternative to hopping above the barrier. Pictorially it is like going through a wall instead of walking around it. Other processes regard a proton transfer in enzymatic reactions or the simultaneous transfer of a proton and an electron from different sites (so-called proton-coupled electron transfer). Perhaps the most surprising outcome is that animal orientation through the earth's magnetic field seems to exploit the magnetic field tuning of the yield of a radical pair reaction. The possibility that biological systems are performing a kind of magnetic resonance experiment to guide their seasonal migration patterns or other navigation is fascinating.

The experiment that is most widely used for highlighting quantum coherence is 2D photon echo spectroscopy. Photon echo allows the measurement of polarization (coherence) dynamics, and the 2D version of this experiment aims at imaging the nondiagonal coherences of a multilevel system. Essentially it is like a real imaging of the density matrix of the system.

The main questions discussed by Fleming regarding quantum coherence in biology and chemistry are (i) Is coherence important? (ii) Of what use is coherence? (iii) Can we detect interference between pathways?

This field of research is attracting a lot of attention because of its relevance to very important and fundamental phenomena and its reach and partially unexplored physics. Time, and the work of many, will possibly provide answers to the questions reported here.

Exciton coupling can be a diagnostic tool for the geometrical arrangement in space of large molecules or proteins. For this purpose, the system under study is functionalized with properly attached chromophores, which interact in different ways depending on the conformation.

Exciton coupling can be a handle for controlling resonance energy transfer and energy migration in aggregates. In photovoltaics, molecular aggregation can be used for tuning the absorption spectrum, for optimal resonant energy transfer and highest efficiency charge injection. In light harvesting systems, exciton resonances are "tuned" by nature by controlling the organization in space of the involved pigments through the protein structure. This leads to optimized energy transport to reactive sites (e.g., the special pair where charge separation takes place).

Finally, the intermolecular interaction V can be the dipole–dipole interaction we mentioned above, or a full Coulomb interaction, between the monomer wavefunctions. The model we briefly sketched is oversimplified because (i) we did not consider intramolecular vibration, (ii) we neglected intermolecular vibrations, and (iii) we neglected the interaction between the dimer and the environment. However, the most striking feature of the molecular exciton, that is, energy delocalization, has been described.

3.2
Vector Model for the Large Aggregate

A pile of flat molecules, face to face as in a card deck, constitutes the extension of the H-dimer we saw above to the so-called H-aggregate. N molecules will give rise to N states in the aggregate. The state with the highest energy is reached when all dipoles oscillate in phase (repulsive interaction), and the lowest when all dipole oscillates pairwise out of phase (attractive interaction). Again, the highest state is optically active with maximum transition dipole moment; the lowest state is dark, with null dipole coupling.

Considering the in-line, tail-to-head arrangement, we obtain a J-aggregate, where the situation is reverse: the lowest state is optically active, the highest state is dark. In more general terms, the two arrangements can also be named by the sign of the intermolecular interaction. $V > 0$ for H type, and $V < 0$ for J type. If k is used for counting the states, the energy spectrum of an aggregate of N molecules is

$$E_k = \varepsilon + 2V\cos\left(\frac{\pi k}{N+1}\right) \tag{3.15}$$

where $k = 1, 2, \ldots, N$. For $k = 1$ and $N \gg 1$ $E_1 \approx \varepsilon + 2V$ is the energy of the optically active state (Figure 3.10). For $N \gg 1$ the exciton bandwidth is $4V$. Each state corresponds to a collective, delocalized excitation. In spectroscopy this has a number of consequences: redistribution of oscillator strength in a few transitions and cooperative spontaneous emission (superradiance) when all the dipoles interfere in phase. The exciton band is narrow because vibrational coupling is strongly depressed by the exciton delocalization (see exciton–phonon coupling renormalization in the next chapter), thus reducing the lineshape to that of a homogeneous zero phonon line. In practice, the $k = 1$ transition collects 80–90% of the total oscillator strength ($\sim N\mu^2$), while the rest go to a few other k-states. The J-aggregate emission from the lower lying coherent excited state is a collective process called *superradiance*. The radiative time is reduced, accordingly, by $1/N$, and this may affect the quantum yield when nonradiative decay becomes slower than

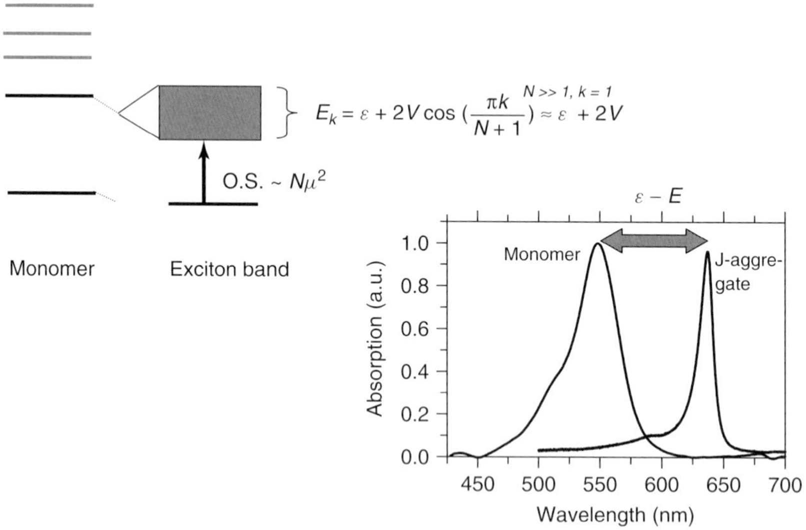

Figure 3.10 J-aggregate for large N, with an example of cyanine dye, with N about 10.

this value. The emission is strong and spectrally confined. The emission linewidth becomes narrower on increasing N, until radiative lifetime dominates dephasing. There are two other emission processes that lead to sharp, intense, emission lines: superfluorescence and amplified spontaneous emission (ASE).

A multichromophore system wherein the transition dipoles add up in phase during a preparation time, before emission takes place, is dubbed superfluorescence. Different from superradiance, superfluorescence is a cooperative effect wherein coherence is formed in the excited state. ASE is a noncoherent cooperative process, where spontaneous emission from any chromophore in the ensemble gets amplified traveling through the system because of stimulated emission. Optical gain is a necessary prerequisite for ASE. Traveling is assured by proper waveguiding conditions in the excited volume of the sample. In ASE neither the excited-state wavefunction nor the emission dipoles are coherent superposition of the components, yet the emitted light that undergoes some spatial filtering during amplification shows directionality (space coherence). This is defined by the geometrical properties of the emitting volume (length and cross section of the excited volume). The typical geometry is a thin, long stripe, with thickness above the cutoff for waveguiding. ASE is similar to lasing but should not be confused with LASER emission. LASER emission is defined by the modes of an optical resonator that contains the active medium and shapes their interaction.

3.3 Interaction Regimes and Vibrational Dynamics

Let V be the excitonic coupling and $\Delta\varepsilon$ the width of the vibronic absorption lineshape of the monomer (Figure 3.11). Strong coupling regime is when, $\frac{2V}{\Delta\varepsilon} \gg 1$.

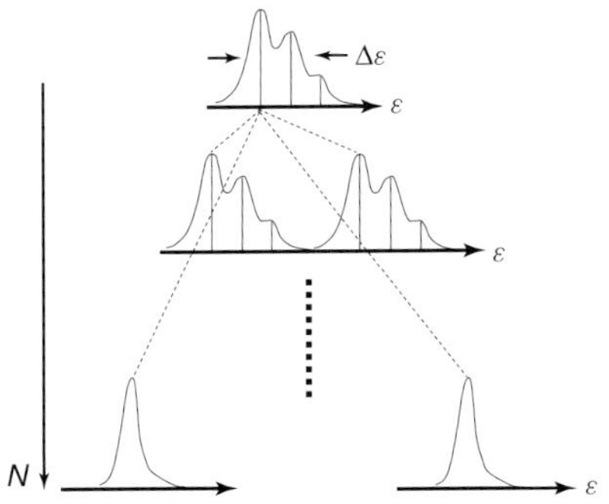

Figure 3.11 Effect of strong coupling in the dimer. At the top is the molecular absorption lineshape. The exciton splitting is specified by the total oscillator strength of the vibronic molecular spectrum. For larger aggregates, from top to bottom ,the size of the exciton increases. The effect on intramolecular vibrations is shown.

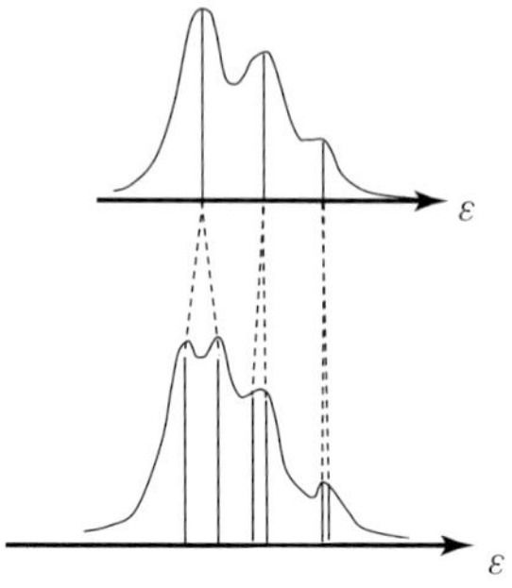

Figure 3.12 The weak coupling regime, when exciton splitting is smaller than vibrational spacing. Each vibronic transition in the molecular spectra specifies the excitonic splitting.

In this case, the adiabatic approximation holds true: the electronic transitions interact to form the exciton band, which is wider than the vibrational spacing, while vibrational replica *follows* adiabatically. The exciton width is $V \approx \frac{\overline{\mu}_0^2}{|R|^3}$, where μ_0 is the electronic dipole moment of the electronic transition worked out onto the electronic wavefunction, $\overline{\mu}_0 = \langle \phi_e(r, R) | \hat{\mu}_e | \phi_g(r, R) \rangle$. The electronic origin is "shifted" by the exciton coupling and degeneracy removed. The vibrational replicas are on top of each exciton transition spreading the oscillator strength. However, the intensity of these replicas is not equal to that in the monomer, because electron–phonon coupling is renormalized within the exciton band. The more the number of molecules coupled in the aggregate the larger the exciton and lower the vibrational replica. In other words, the Huang-Ryss factor depends on the exciton coupling strength. For $N \gg 1$ and strong coupling, the vibrational replica disappears. The reason is that on spreading of the electronic wavefunction onto the aggregate the local amplitude of excitation becomes smaller, reducing the effect of distortion in the nuclear configuration of the molecules.

For weak coupling $\frac{2V}{\Delta\varepsilon} \ll 1$, each vibronic transition interacts separately, because the adiabatic approximation breaks down. The exciton splitting of the 0–n vibronic transition is roughly $V \approx \frac{\overline{\mu}_0^2 F_{0n}}{|R|^3}$, where F_{0n} is the FC factor of the molecular transition (Figure 3.12).

The intermediate coupling regime, when $\Delta\varepsilon \approx 2\ V$, is the most frequent in nature and the most difficult to be defined. Figure 3.13 suggests a possible scenario.

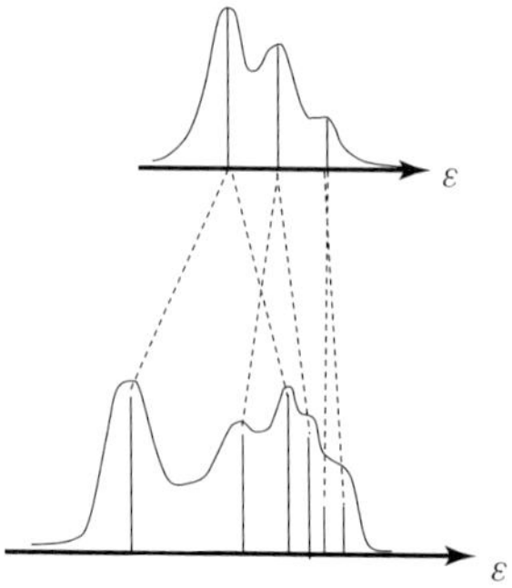

Figure 3.13 Intermediate coupling. Here things cannot be simplified, and the aggregate spectrum is somewhat a mixture of both Figures 3.14 and 3.15.

Very weak coupling is when the electronic wavefunctions and transition energy are unchanged, yet some resonance interaction may still have an effect in the energy propagation, leading to a slow transfer rate. In this case, the interaction energy is evaluated to the second order in the dipole–dipole interaction, leading to a distance dependence $V \sim R^{-6}$ typical of Foerster energy transfer. The energy transfer process is not coherent and can be time resolved (i.e., the propagation of energy among states is an observable). The typical rate is time dependent, or dispersive, $\gamma(\tau) \sim t^{-1/2}$. Owing to its importance in soft matter this is discussed separately.

3.4 Quick View on Aggregate Multiexciton States

When exciton coupling is strong, the aggregate loses all the molecular properties and behaves as a new system. Its elementary excited states are excitons in the aggregate band structure, specified by the one-exciton band, the two-exciton band, and so on. The one-exciton band corresponds to exciting one molecule in the aggregate, and it accommodates N states, one state per molecule. The two-exciton band regards the excitation of two molecules in the aggregate and contains $N(N-1)/2$ states, because this is the number of possible ways to prepare such a state, excluding a doubly excited single molecule and accounting for identical molecules (permutation of the excited pair). The three exciton band will have $N(N-1)(N-2)/3!$, and so on.

Let us consider a J-aggregate. Following excitation in the one-exciton band, after relaxation, the lowest states will be occupied. Transmission difference spectrum, as measured in pump probe, will show photobleaching (PB) and stimulated emission (SE) from the lowest one-exciton band state ($k = 1$), that is, strongly allowed, and excited-state absorption (photoinduced absorption (PA)), from the one-exciton band to the next two-exciton band. PA will undergo a blueshift with respect to PB, giving rise to a typical lineshape as shown in Figure 3.14. The energy separation between

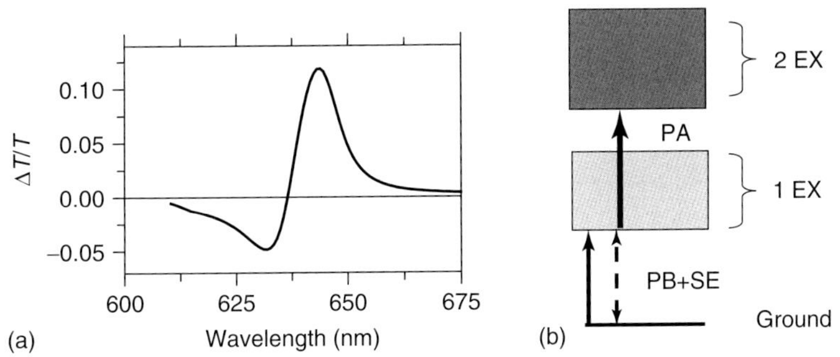

Figure 3.14 (a) Pump probe spectrum of J-aggregate. (b) The exciton band structure of the aggregate.

PA and PB can be used to estimate the aggregate size and coupling strength, according to

$$\Delta\varepsilon = \hbar\omega_{PB} - \hbar\omega_{PA} = 2V\left[\cos\left(\frac{\pi}{N+1}\right) - \cos\left(\frac{2\pi}{N+1}\right)\right] \cong -\frac{3\pi^2 V}{(N+1)^2} \qquad (3.16)$$

where we assume $N \gg 1$ to obtain the last equation. The energy shift between the monomer and aggregate, $V = \frac{E-\varepsilon}{2}$, measured by the linear absorption spectra, can be used to estimate V, which can then be introduced into Eq. (3.16) to obtain N. Suppose the molecular peak is at $E = 2.3$ eV and the J-aggregate resonance is at $\varepsilon = 1.9$ eV, then $V = -0.2$ eV, and assuming $\Delta\varepsilon = -0.01$ eV, $N = \pi\sqrt{\frac{3V}{\Delta\varepsilon}} - 1 \approx 23$.

For the dimer (with $k = 1, 2$) the PA transition connects the low-energy dimer split-off state with the doubly excited state. The energy of this state is simply twice that of the single molecule excitation, according to Heitler–London approximation, assuming $2E \gg V$. The PA transition has energy $2E - (E - |V|) = E + |V|$ and the separation between the PA and the PB peaks is $2V$.

3.5 The Role of Disorder

A distribution of site energy due to disorder, $\varepsilon_n = \varepsilon + \delta\varepsilon_n$, may affect the exciton structure till its full destruction and consequent wavefunction localization, as schematically drawn in Figure 3.15. Assuming that the energy distribution due to disorder is described by a function $P_\sigma(\varepsilon)$ of width σ we can distinguish three

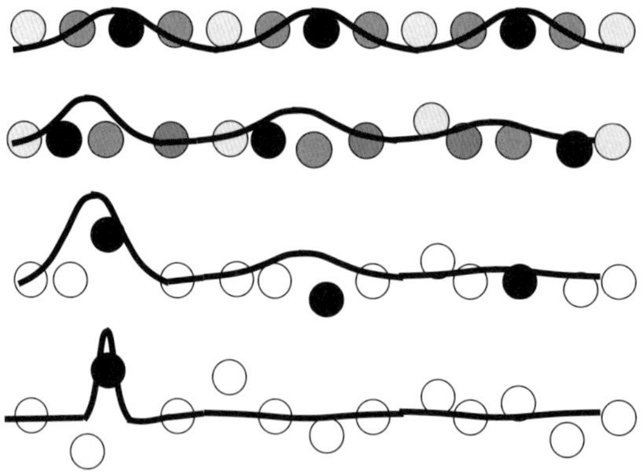

Figure 3.15 A schematic of the exciton localization phenomenon due to disorder. Initially (first raw) the probability amplitude for molecular excitation is evenly distributed on the aggregate. Weak disorder (second raw) redistributes such probability, affecting the regular pattern to a minor extent. In the third raw, because of strong disorder the probability amplitude is almost localized, and finally it is only on one site when the disorder is very strong. In the fourth raw, the exciton state is fully destroyed.

Weak (no state mixing)

$\sigma \ll E_2 - E_1$ } $\text{FWHM} \approx \sigma / \sqrt{N}$ Exciton energy fluctuations

Medium (state mixing)

$\dfrac{\sigma}{V}$ } $N \rightarrow N_D$

Exciton localization

$(N \gg N_D \gg 1)$

Strong (no exciton)

$\sigma \gg V$ } $\text{FWHM} \approx \sigma$ Individual molecule limit

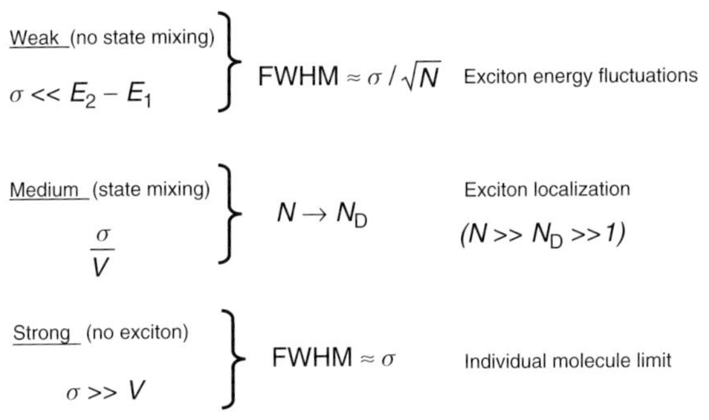

Figure 3.16 The three regimes of molecular disorder and their effect on the aggregate exciton state.

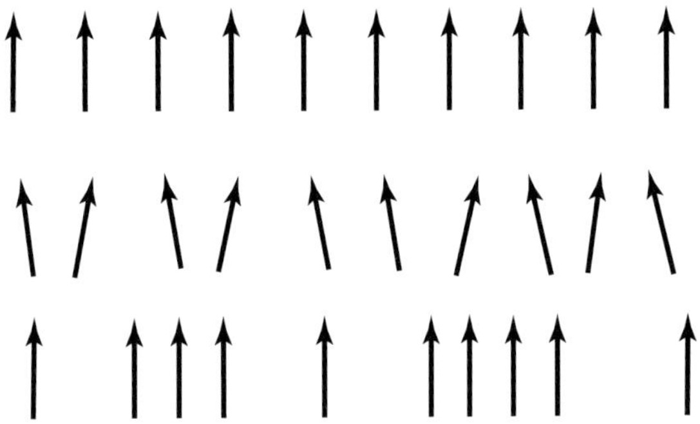

Figure 3.17 Different kinds of positional disorder.

regimes, as described in Figure 3.16. In case of weak disorder, the disorder-induced interactions are too small to induce effective state mixing. This will cause small exciton energy fluctuations and broadening of the exciton resonance without breaking the coherent state. Intermediate disorder will cause exciton localization because of state mixing. This will reduce the number of molecules effectively involved in the collective excitation. The coherent state is then extended to a number of molecules smaller than the total number of molecules in the aggregate, and the coherence length is $L_C = N_D a$, where a is the intermolecular distance. This is the most common situation in the real world, when an aggregate of size N can actually support a coherent collective excitation onto a smaller number of molecules, $N_D < N$.

Finally, when disorder fluctuations are larger than the exciton coupling energy, the collective coherent state cannot be established. The observed energy band will be fully incoherent because of a distribution in energy of localized molecular states.

Disorder may be caused by fluctuations in the molecular position and orientation in the aggregate, as shown in Figure 3.17, thus changing the intermolecular interaction. As an example, we look at H-aggregate. Let us assume that a random tilting angle is introduced, leading to oblique dipoles. This changes the selection rule and activates the low-energy transition, as discussed above for oblique dipole dimer. Another effect is breaking down of the aggregate into smaller aggregates, when some molecules are closer than others in pairs, trimers, or larger groups. This leads to wavefunction localization and transition energy broadening, according to the n-mers distribution.

4
Excited States in Solids

4.1
On the Origin of Bands in Solids

Energy bands are characteristic of aggregates of fundamental units (atoms or molecules). The discrete energy levels of the original unit get distributed into a continuum of states when interacting in the aggregate/solid, that is, on reducing the interunit distance. Two very different situations can occur: coherent and incoherent energy bands.

Let us consider the electronic absorption line within the two-level model for an isolated atom or molecule, Figure 4.1. The full width at half maximum (FWHM) is given by

$\Delta\omega_{\mathrm{FWHM}} = 2\Gamma = \frac{2}{T_2}$, and the homogeneous lineshape in the spectrum is Lorentzian:

$$A(\omega) = C\omega \cdot \mu^2 \frac{\Gamma}{(\omega - \omega_0)^2 + \Gamma^2} \tag{4.1}$$

This describes an electronic transition without vibronic coupling. Electronic dephasing and linewidth comes from the interaction with the environment. In an ensemble of noninteracting molecules this equation is still good, once linewidth is ascribed to homogeneous broadening. Homogeneous broadening means that each molecule in the ensemble encounters the same environment and perturbations, and a unique distribution function of the Lorentzian shape describes the energy fluctuation. When each molecule in the ensemble is, in addition, subjected to a local perturbation that is different for each site, the broadening acquires an inhomogeneous contribution, and the over all lineshape becomes the convolution of the homogeneous lineshape with a distribution lineshape (typically Gaussian). This is called *Voigt profile* (Figure 4.1):

$$A_I(\omega) = P_\sigma(\omega) \oplus A(\omega) \approx P_\sigma(\omega) \tag{4.2}$$

Here, P_σ is the inhomogeneous distribution. The width of the absorption line will be larger than that in the homogeneous case. In most situations (molecules in solution or solid at room temperature), disorder dominates the lineshape, and the spectral width actually reflects the variance of the inhomogeneous distribution (like if the homogeneous line were a δ-function in frequency). In this case, the

The Photophysics behind Photovoltaics and Photonics, First Edition. Guglielmo Lanzani.

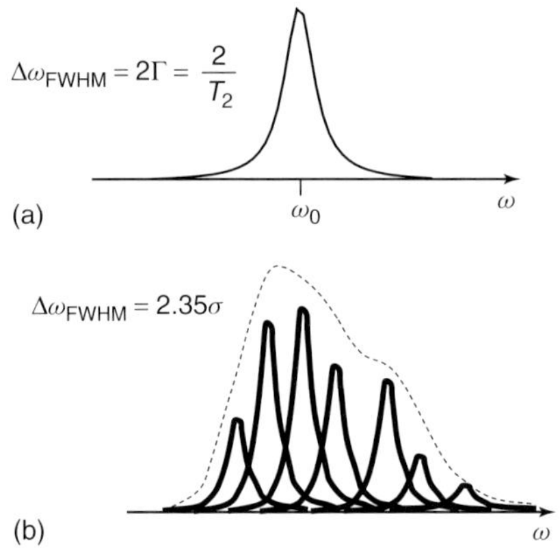

Figure 4.1 (a) Homogeneous linewidth. (b) Inhomogeneous linewidth.

FWHM is described by $\Delta\omega_{FWHM} = 2.35\sigma$, where σ is the standard deviation of the distribution. For instance, in a molecular solid, disorder induces a Gaussian density of states (DOS) such as

$$g(E) = \frac{1}{\sqrt{2\pi}\sigma} \exp\left(\frac{E^2}{2\sigma^2}\right) \tag{4.3}$$

Such a distribution function may well describe both the DOS of the neutral and charged states. There is no direct experimental proof that the distribution is Gaussian. However, it is known that inhomogeneous broadening of optical spectra is in most cases of Gaussian shape. The reason is that the lattice polarization energy of an excited molecule depends on many internal coordinates, each varying randomly. In these conditions on can apply, the central limit theorem that predicts Gaussian envelope function regardless of the specific type of interaction.

Disorder induces the formation of bands in molecular solids or any other amorphous material with weak interunit interactions. These bands correspond to different energies in different locations in real space. Once an excited-state population is generated by any external excitation with a characteristic initial DOS occupation probability, energy can migrate within the inhomogeneous DOS, changing the occupation distribution (typically there will be an overall downshift of the average energy and a broadening of the distribution shape, tailing to lower energy). This occurs by incoherent energy hopping between sites in real space. Concomitantly after electronic energy relaxation excess energy is dissipated by intramolecular vibrational redistribution.

Many experimental results reflect the presence of a disorder-induced band, for instance:

1) Absorption spectral lines are broader in aggregates with respect to isolated molecules;
2) Emission spectra can be narrower because of spectral migration to lower states before emission (a sort of intermolecular Kasha rule). When this happens, a sizable Stoke shift is observed between the absorption and emission maxima. When this does not happen, energy is reemitted by the absorbing site, a

situation known as *resonant emission*. Typically this occurs below a certain threshold energy called *localization threshold*.

3) The lifetime of the excited state is not well defined but is described by a distribution of values within a range. Correspondingly, the decay kinetics of the excited-state population is not monoexponential.
4) Charge transport is dispersive, that is, there is a distribution of hopping time constants for carriers and transport is not Gaussian. A packet or sheet of carriers will spread in space and time without a well-defined transfer time.

Energy bands in aggregate/solids may arise also from a very different phenomenon: the interaction between the components (atoms or molecules). This gives rise to coherent energy bands. For infinitely separated atoms, each electronic state is N-degenerate, where N is the number of atoms in the lattice. On reducing the interatomic distance, the degeneracy is lifted and the N-states spread into a band, forming a quasi continuum for large enough N. This band is, however, very different from that discussed before. Each state is fully delocalized in space, while disorder induces a localized state with different energy in each position. Propagation of energy or charge is coherent and wavelike in nature, whereas it is by hopping in the disorder case.

Each state in the band represents a collective excitation of the whole crystal with wavefunction fully delocalized over the lattice or ensemble. In the presence of a periodic lattice, the quantum number that characterizes each state in the band is a vector $\bar{k}$ in the reciprocal space of the lattice. Propagation is described by semiclassical states built by the coherent superposition of band states, known as *wavepackets*. This is discussed in detail in the section below. Energy depends on the wavevector $\bar{k}$ according to a dispersion law, $E(\bar{k})$. In n-D $\bar{k}$ is an n-component wavevector, and dispersion might be different along each direction.

Now we briefly describe the equation of motion of an electron in an energy band. The electron is described by a wavepacket made by the superposition of delocalized states within a certain range Δk around a particular wavevector k. (We adopt here a simple 1D picture dropping vector notation.) The group velocity of the wavepacket is $v_g = \frac{\partial \omega}{\partial k} = \frac{1}{\hbar}\frac{\partial E}{\partial k}$. The effect of the crystal on the electron motion is contained in the dispersion law $E(k)$. At the bottom of the band, the derivative is null and the wavepacket immobilized. A force F applied to the electron can change its state according to $\hbar\frac{\partial k}{\partial t} = F$, where the quantity $\hbar k$ plays the role of momentum in a formal equivalent of Newton's second law. Thus we can differentiate in time the group velocity equation to obtain

$$\frac{\partial v_g}{\partial t} = \frac{1}{\hbar}\frac{\partial^2 E}{\partial k \partial t} = \frac{1}{\hbar}\frac{\partial^2 E}{\partial k^2}\frac{dk}{dt} = \left(\frac{1}{\hbar^2}\frac{\partial^2 E}{\partial k^2}\right) F \tag{4.4}$$

If we identify $\left(\frac{1}{\hbar^2}\frac{\partial^2 E}{\partial k^2}\right)$ with a mass, Eq. (4.4) assumes the form of Newton's second law $F = m^*\frac{\partial v_g}{\partial t}$, where we define effective mass m^* by $m^* = \left(\frac{1}{\hbar^2}\frac{\partial^2 E}{\partial k^2}\right)$.

By the assumption that $\hbar k$ behaves like momentum in the Newton's second law for the electron in the conduction band, we can define the effective mass. One may

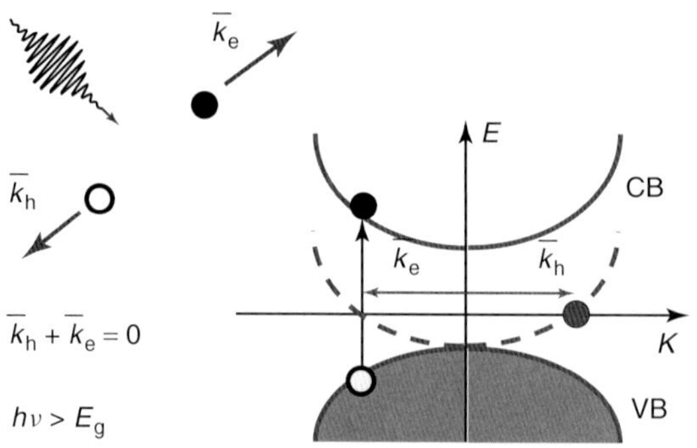

Figure 4.2 Electron and hole photogeneration. The dashed line shows the hole band that simulates the dynamics of a hole, constructed by inversion of the valence band in the vertex. The wavevector and energy of the hole are equal but opposite in sign to the wavevector and energy of the empty electron orbital in the valence band.

ask if this is general and justified. The response is positive. A rigorous derivation of $\hbar \mathrm{d}k/\mathrm{d}t = F$ can be found in many solid state physics books.

In semiconductors, at $T = 0$, the fully occupied highest energy band is the valence band; the first empty band, separated by the energy gap, is the conduction band. Photon absorption promotes an electron from the valence band to the conduction band (Figure 4.2). The electron has a wavevector $\bar{k}_\mathrm{e}$ and charge $-e$. The vacant orbital left in the valence band is modeled as a positive particle, named hole. The filled band has total wavevector $\sum \bar{k} = 0$. If the electron is missing from a state of wavevector $\bar{k}_\mathrm{e}$, the wavevector of the vacant state is $-\bar{k}_\mathrm{e}$. This is the wavevector of the hole $\bar{k}_\mathrm{h} = -\bar{k}_\mathrm{e}$. The total wavevector of the e–h pair is $\bar{k}_\mathrm{h} + \bar{k}_\mathrm{e} = 0$, according to the selection rule based on photon momentum conservation. The energy of the hole is $E(\bar{k}_\mathrm{h}) = -E(\bar{k}_\mathrm{e})$, that is, the energy of the missing electron in the valence band. Let the zero of energy of the valence band be on top of the band. The lower in the band the missing electron lies the higher the energy of the system. The opposite sign of energy takes into account that more work is needed to remove an electron from a deep state than from a shallow one. Because of the reverse sign in energy, the proper representation of the hole band is upside down with respect to the valence band, that is, it has the same curvature of the conduction band. According to the band picture in Figure 4.2, the effective mass of both the electron and the hole is positive. The hole has the velocity of the missing electron, that is, opposite to that of the electron in the conduction band. Within this picture, photon absorption in a semiconductor gives rise to pair generation, with conservation of charge, momentum, and energy. However, with respect to the same process in vacuum, mass is conserved and the crystal does not weight differently when electron and hole are generated.

The two particles are formed with opposite momentum and thus depart from each other in space.

The excess photon energy $\varepsilon = h\nu - E_\mathrm{gap} \approx \frac{\hbar k^2}{2m_\mathrm{e}^*} = E_\mathrm{e}(k_\mathrm{e})$ goes essentially into the kinetic energy of the electron because usually $m_\mathrm{h}^* \gg m_\mathrm{e}^*$.

The DOS for each energy band is of particular relevance for photophysics and transport. The number of states between E and $E + \mathrm{d}E$ per unit volume is $g(E)\mathrm{d}E$, where $g(E)$ is the DOS distribution function. Because each state $E(k)$ corresponds to a value k, one can count how many k-values are allowed in the elementary volume of the k-space corresponding to the energy range, $E + \mathrm{d}E$. In 3D the elementary volume is $\mathrm{d}\bar{k} = 4\pi k^2 \mathrm{d}k$. The number of allowed k-values is $\frac{\mathrm{d}\bar{k}}{\nu_k} \propto \mathrm{d}\bar{k}$ where ν_k is the volume per k-value. In the isotropic effective mass approximation $E(k) = \frac{\hbar^2 k^2}{2m^*}$, so $k^2 \propto E$ and $\mathrm{d}k \propto \frac{1}{\sqrt{E}}\mathrm{d}E$ giving $g(E) \propto \mathrm{d}\bar{k} \propto \sqrt{E}$.

In 2D $\mathrm{d}\bar{k} = 2\pi k\mathrm{d}k$ and $g(E)$ is constant.
In 1D $\mathrm{d}\bar{k} = \mathrm{d}k$ and $g(E) \propto \frac{1}{\sqrt{E}}$.

A different dispersion law leads to a different DOS. For instance, in graphene or metallic carbon nanotubes $E(k) \propto k$, a feature that is associated with massless Dirac Fermion (because it behaves like photon dispersion), and for 2D $g(E) \propto E$, while for 1D $g(E)$ is constant.

In band semiconductors, absorption is due to valence band to conduction band transitions, which involve Bloch states. In general, the dipole moment of each transition depends on the k-vector; however assuming spherical symmetry allows dropping the vector character of k, introducing an average dipole, and a further approximation is that this dipole is constant over the range of allowed k. Under this approximation, the absorption spectrum maps the density of the initial and final states, coupled by the transition, called *joint density of states* J_{CV}. Energy conservation requires that all and only those states such that their energy separation $E_{\mathrm{cv}}(k) = \hbar\omega$ are involved in optical transition at energy $\hbar\omega$. Assuming parabolic bands $E_{\mathrm{cv}}(k) = E_{\mathrm{gap}} + \frac{\hbar k^2}{2\mu_{\mathrm{eh}}}$, where μ_{eh} is the reduced mass of the electron–hole pair defined as $\mu_{\mathrm{eh}} = \left(\frac{1}{m_{\mathrm{e}}} + \frac{1}{m_{\mathrm{h}}}\right)^{-1}$. This allows transformation of the calculation of J_{CV} from k-space to energy $\int \mathrm{d}\bar{k}\delta(E_{\mathrm{cv}} - \hbar\omega) = \frac{4\pi}{\hbar^3}\mu_{\mathrm{eh}} \int [2\mu_{\mathrm{eh}}(E_{\mathrm{cv}} - E_{\mathrm{gap}})]^{1/2}\delta(E_{\mathrm{cv}} - \hbar\omega)\mathrm{d}E_{\mathrm{cv}}$, which implies a behavior of the absorption coefficient $\alpha \propto \left(\hbar\omega - E_{\mathrm{gap}}\right)^{1/2}$.

There exists an oscillator sum rule also for solids, which can be written as $\int_0^\infty \omega \,\mathrm{Im}\,\chi(\omega)\mathrm{d}\omega = \frac{1}{2}\pi\omega_{\mathrm{P}}^2$, where $\omega_{\mathrm{P}}^2 = \frac{4\pi Ne^2}{m_{\mathrm{e}}}$ is the plasma frequency, or alternatively as $\sum_j f_{jh}(k) = 1 - \frac{m_{\mathrm{e}}}{m_{\mathrm{h}}^*(k)}$. Here, $f_{jh}(k)$ is the oscillator strength for the transition from the state at k in filled band h to state k in band j. For flat, narrow bands the effective mass m_h^* is very large, and the sum rule tends to the molecular expression. However, for parabolic bands (free electrons) with $\frac{m_{\mathrm{e}}}{m_{\mathrm{h}}^*(k)} \approx 1$, the total oscillator strength can be very small.

Concluding, the very existence of sum rules for the oscillator strength tells you two important things: (i) the absorption strength of a material is limited by its electronic density (number of available electronic transitions) and (ii) on population redistribution, new transitions are activated (photoinduced absorption) and consequently others should be reduced (photoinduced bleaching).

4.2 Excitons

Excitons are quasiparticles that represent a collective excited state of an ensemble of atoms or molecules in an aggregate or in crystals. A quasiparticle is a semiclassical state that is represented by a wavepacket for which we can define mass and speed. Excitons are neutral (currentless) wavelike excitations. Each exciton state is monoenergetic, like a monochromatic component in a pulse of light. Real excitations are never monochromatic but possess a certain energy bandwidth, determined by the generation process. The coherent superposition of excitonic states gives rise to the wavepacket. An excitonic wavepacket transports energy according to the group velocity and the effective mass defined by the energy dispersion curve.

Excitons affect the optical properties of the aggregate or crystal. Their properties, such as spatial extension, speed of propagation, optical transitions, and lifetime, depend on the type of interaction between the atoms or molecules in the ensemble, on their spatial arrangement and relative orientation, and also on the symmetry of the crystal lattice. Excitons can be characterized by the binding energy, that is, the energy required to separate them into a pair of free charge carriers; however, this classification is not general and depends on the type of interaction and on the dimensionality of the crystal.

There are two extreme cases: the Frenkel exciton regards solids that are made by weakly interacting units, for example, molecular crystals. The Frenkel exciton is the general case of the molecular exciton described previously according to Kasha. The wavefunction amplitude squared is associated with the probability of finding the excited state in the lattice. The excited state carried around by the exciton is molecular-like, localized on a single molecule. How much of this excitation each molecule carries is described by the exciton wavefunction. Frenkel exciton wavepackets are narrow in space, because of the rather flat dispersion curve (large effective mass) in k-space.

The Wannier–Mott (W–M) exciton regards solids that are made by tightly bound atoms, for example, covalent solids. This exciton describes the propagation of an electron and hole pair. According to the classic picture, electron and hole orbit around each other at a distance (exciton radius) much larger than the lattice constant. This system resembles a hydrogen atom where the proton is now a positron. The exciton wavefuction describes two observables: the electron–hole motion and the center of mass motion. Propagation of the center of mass is described by a wavepacket, with quasi continuum energy dispersion in k-space. The electron–hole motion is described by hydrogenoid wavefunctions associated with a discrete spectrum of energy states.

Both Frenkel and W–M exciton states (with specified energy and quasi momentum) are stationary waves delocalized on the whole crystal. The coherent superposition of such states, however, is a wavepacket with finite spread in real (ΔX) and reciprocal (ΔK) space. This is the semiclassical state. We now describe in more detail the two extreme cases.

4.2.1
Frenkel Exciton

In order to introduce the concept of the molecular exciton, we use the simple 1D example. We consider N molecule in a linear lattice with spacing "a." We consider a molecular solid where the intermolecular coupling is much weaker then the intramolecular coupling. Following excitation, for instance by light, because of the intermolecular interaction, the molecular excited state becomes delocalized on the whole crystal or aggregate. It is a collective state described by a Bloch wavefunction and characterized by a quasi momentum k, which is a good quantum number. The exciton wavefunction describes the probability of finding the molecular excited state in a particular lattice location. Because the intermolecular interaction is small, the ground-state wavefunction of the system can be simply expressed as the product of the N molecular wavefunctions (Figure 4.3).

$$\Psi_g = u_1 u_2 \ldots u_{N-1} u_N \tag{4.5}$$

If the jth molecule is excited, and without considering intermolecular coupling, the excited-state wavefunction will be

$$\varphi_j = u_1 u_2 \ldots u_{j-1} v_j u_{j+1} \ldots u_{N-1} u_N \tag{4.6}$$

where "v_j" represents the molecular excited state. Without intermolecular interactions, there are N degenerate states such as that in Eq. (4.6), with energy ε, solution of the molecular Schrödinger equation $H\varphi_j = \varepsilon_j \varphi_j$, where $\varepsilon_j = \varepsilon \quad \forall j$.

If there is intermolecular interaction, the Schrödinger equation for the crystal state, assuming nearest neighbor coupling only, is $H\varphi_j = \varepsilon_j \varphi_j + V\left(\varphi_{j-1} + \varphi_{j+1}\right)$.

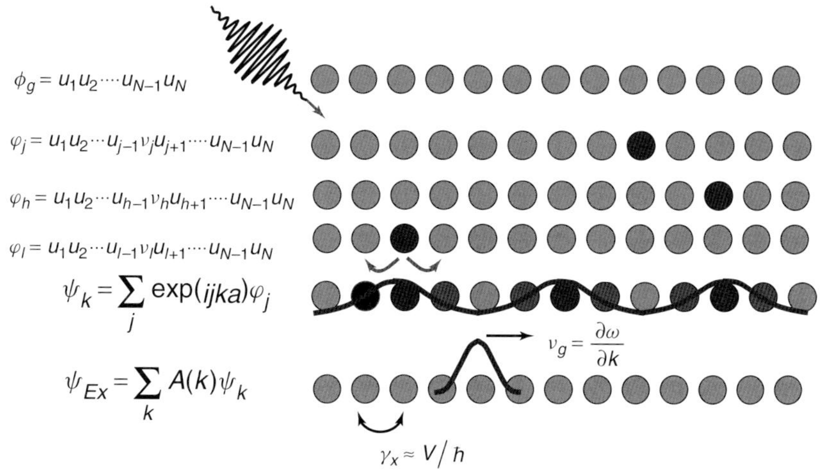

Figure 4.3 The construction of the Frenkel exciton wavefunction and its propagation by coherent motion. The wavepacket has a spread Δk in k-space and Δx in real space with $\Delta k \Delta x \leq \hbar$.

"V" is the intermolecular interaction energy, and it can be of the same origin as discussed for the molecular exciton. In other words, V represents the rate of energy transfer from molecule j to molecule $j \pm 1$. Now each state φ_j is no more an eigenvalue of the energy matrix. Diagonalization leads to new energies and a new basis set, with wavefunctions expressed by linear combination of the φ_j according to

$$\Psi_K = \sum_j e^{ijKa} \varphi_j \tag{4.7}$$

These are coherent superpositions of the original product wavefunctions, with a well-determined phase that depends on the quantum number K. $\hbar K$ plays the role of the exciton quasi momentum.

By introducing wavefunction (Eq. (4.7)) into the Schrödinger equation one gets

$$H\Psi_K = \sum_j e^{ijKa} \left[\varepsilon + V\left(e^{-iKa} + e^{iKa}\right)\right] \varphi_j = (\varepsilon + 2V \cos Ka)\, \Psi_K = E(K)\Psi_K \tag{4.8}$$

where K can be considered as a quasi continuum variable spanning the first Brillouin zone. The intermolecular coupling V removes the degeneracy, forming a band of states whose width in energy is $4V$, as seen in Figure 4.4. Momentum conservation and the fact that photons have a very small wavevector imply that only excitons with $K \approx 0$ are optically allowed. For $V > 0$, the absorption peak is shifted to higher energy with respect to the molecular transition, as in the H-aggregate, and according to Kasha crystal emission is forbidden. For $V < 0$, there is a redshift of the exciton peak, as in the J-aggregate, and emission is allowed. The formation of excitons also implies the renormalization of the electron–phonon coupling, as discussed for the aggregate. As a result, the vibrational progression due to intramolecular vibration is usually reduced, and for strong coupling it is fully suppressed. Excitons can, however, show coupling to intermolecular vibrations (lattice modes).

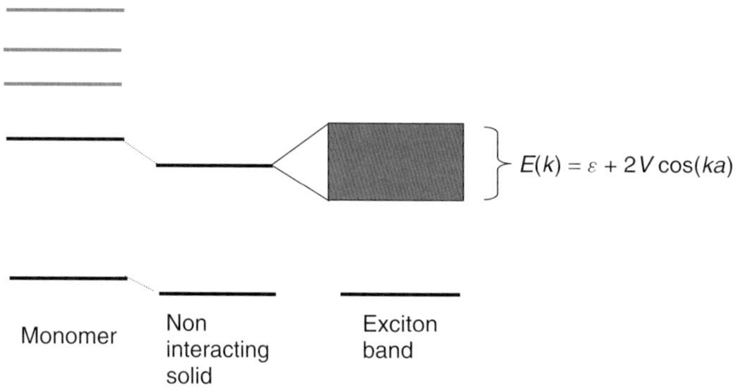

Figure 4.4 In the interacting solid, degeneracy is removed. The process is shown only for the lowest excited state. Each molecular state, however, will split into a band of different widths, depending on the properties of the state.

For $Ka \ll 1$, we can estimate the exciton effective mass in terms of the bandwidth by expansion of the cosine term ($\cos(ka) \cong 1 - \frac{1}{2}k^2a^2$):

$$E(K) = \varepsilon + 2V\cos Ka \approx \varepsilon + 2V - VK^2a^2 = \varepsilon + 2V + \frac{\hbar^2 K^2}{2M_X^*} \tag{4.9}$$

from which

$$M_X^* = \hbar^2 \left(\frac{\partial E}{\partial K}\right)^{-1} = -\frac{\hbar^2}{2Va^2} \tag{4.10}$$

Assuming $a \approx 0.5$ nm, $V = 0.1$ eV, we find $|M_X^*| \approx 2 \times 10^{-29}$ kg. This is about 20 times larger than the free electron mass, demonstrating that Frenkel excitons are "bulky" particles. The maximum speed under coherent motion is given by the group velocity $v_X(K = -\frac{\pi}{2a}) = \frac{1}{\hbar}\frac{\partial E}{\partial K} = \frac{2Va}{\hbar} \cong 3 \times 10^4$ m/s. Figure 4.5 provides a simple picture for the exciton wavepacket.

In the Davydov picture, the delocalized collective exciton state is established by the coherent transfer of energy between the molecules at rate $\gamma_X \approx \frac{V}{\hbar} = 10^{-14}\ \mathrm{s}^{-1}$. This energy transfer should not be confused with the exciton propagation as described above by v_X. The two processes describe two different coherences: γ_X is associated with the coherent superposition of molecular states and the energy delocalization in one exciton state; v_X is associated with the wavelike propagation of the coherent superposition of exciton states forming the wavepacket. Such propagation fulfills K conservation, till a scattering event changes K, or eventually fully destroys the wavepacket. In this event, the exciton becomes localized and loses all its properties, essentially becoming a molecular excited state. Because this state can still propagate, yet by incoherent hopping (for instance, due to Förster coupling), it is sometime called "*exciton*" as well. Strictly speaking, however, this is no more an exciton state because it has lost all the quasiparticle characteristics. The typical rate of energy transfer by incoherent hopping is $\gamma_M \ll 10^{-13}\ \mathrm{s}^{-1}$.

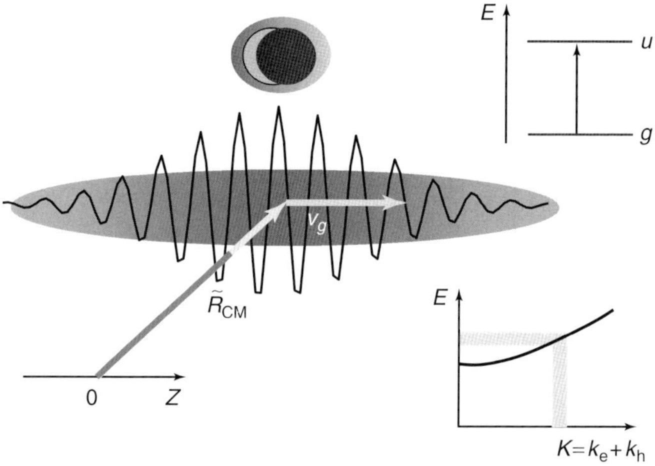

Figure 4.5 The Frenkel exciton wavepacket associated with a molecular excitation g–u.

Coherent, excitonic transport is much more efficient than incoherent hopping, and it can have important role in photovoltaic, see Box 4.1. Excitons have a mean free path, $l_X \gg a$. For ballistic transport, neglecting scattering, the distance traveled during exciton lifetime $\tau_X \approx 1$ ns and using approximate estimates as above is $l_X = 3 \times 10^{-5}$ m. This is a huge distance, only possible in an ideal crystal. In the weak scattering regime, an exciton propagates coherently with well-defined quasi wavevector (center value of the wavepacket distribution) in between scattering events. On scattering, the quasi wavevector will change from K_X to a different value K'_X. In 1D transport, the diffusion constant for the weak scattering regime can be defined by assuming thermal equilibrium and

Box 4.1: Quantum Coherence in Photovoltaics

The coherent or incoherent nature of the excited states dictates the efficiency of energy coupling and transport. This generally applies to energy as well charge transport. The use of crystals in molecular photovoltaic could bring some advantages, for instance, better transport, but it is rarely adopted because of practical limitations in growing defect-free crystals.

Aggregates supporting excitonic excitations could be implemented as light harvesting units in new photovoltaic architectures. In principle, this would allow strong absorption at the exciton resonance and efficient transport through the quantum coherence path. In practice this has not yet been exploited.

A different topic is the question of quantum coherence in polymer films. Following the recent debate on the role of quantum effects in biology, and the indication of long-lasting quantum coherence in complex biological systems, there is an equivalent idea that a coherent state could be involved in the initial light absorption in spite of the reduced order of polymer films. Small excitons (H-aggregates, for instance) can explain optical spectra, as demonstrated by Frank Spano and Jenny Clark. They could also be involved in the initial energy transfer. Even more far off the conventional picture, the initial photoexcitation in polymer could be a free pair of charges, which only later recombines into an exciton state. This short-lived free charge pair could lead to ultrafast charge separation, according to the SSH model (see Heeger *et al.* (1988)).

The existence of a possible hybrid Frenkel–Wannier–Mott exciton at the interface of organic and inorganic semiconductors has been discussed by Vladimir Agranovich. The exotic nature of this state would take the better of both excitations, for instance, high oscillator strength (Frenkel) and excellent transport properties and easy dissociation (W–M). However, experiments are still struggling to find solid evidence for that.

Excitons are considered the primary photoexcitations in carbon nanotubes, which may play a role in photovoltaics. Polycrystalline films are made by an ensemble of nanocrystals of different sizes and orientations, whose excitations are well described by the exciton model. Pentacene, whose singlet exciton can undergo efficient singlet fission, is a study case for Frenkel exciton formation.

using the equipartition theorem: $\frac{1}{2}M^*v_X^2 = \frac{1}{2}M^*\frac{D_X T_X}{T_X^2} = \frac{1}{2}K_B T$. Here T_X is the coherence time between scattering events that change the exciton wavevector. Assuming a scattering time $T_X \approx 0.1$ ns and exciton mass $M^* = 10\, m_0$ at room temperature, we obtain $D_X = \frac{K_B T}{M^*} T_X \approx 5 \times 10^{-2}$ cm^2/s. Using this value, the corresponding mean free path is $l_X = \sqrt{D\tau_X} \approx 70$ nm. In the strong scattering regime, the exciton wavevector is not well defined, and the transport process switches to incoherent hopping, associated with a diffusion constant $D = \gamma_M a^2 \approx 10^{-4}$ cm^2/s (assuming $\gamma_M = 10^{-12}$ s^{-1} $a = 0.1$ nm). The mean free path $l_M = \sqrt{D\tau_X} \approx 3$ nm $\ll l_X$.

The two processes of energy transfer can be distinguished by their different temperature dependence. At low T the coherent transport is enhanced because scattering is reduced, thus the coherent state lives longer and l_X is larger. At high T the hopping rate is larger, for two reasons: (i) thermal broadening induces larger spectral overlap between neighboring molecules, which in turn enhances dipole coupling and thus energy transfer, and (ii) thermal-activated hops become more likely.

The energy diffusion constant can be measured in several ways:

1) by measuring the photoinduced polarization anisotropy decay in a pump probe experiment;
2) by measuring the quenching kinetics of the excited states at impurities, once the concentration and thus the average distance of the quenchers is known.
3) by measuring the quenching rate versus distance from a quenching interface.

In both (1) and (2), by varying the distance of the quenchers one can estimate the mean free path. By comparison with the lattice constant one can make an educated guess on the type of transport that is taking place and thus estimate the diffusion constant. Experimental values can then be compared to theoretical prediction according to the dispersion law, for coherent transport, or the Förster rate and lifetime for incoherent transport.

The scattering or coherence time T_X can be estimated from the exciton resonance linewidth, by assuming $T_2 = 2T_X$ because optical dephasing is caused by the same scattering events responsible for decoherence and localization of the exciton wavefunction.

4.2.2
The Wannier–Mott Exciton

Single-particle picture assumes that vertical transitions from valence (hole) to conduction (electron) band leave all other filled states unchanged. In other words, Bloch states take into account the electron–ion interaction but not the electron–electron interaction. In reality, when an electron is pushed into the conduction band, all other states in the crystal also get affected. This is a complicated multiparticle problem, whose first approximation is the two-particle system, as described by the W–M picture.

According to the band-to-band transition described above, we assume that following photoexcitation an electron–hole pair is generated, with total k-vector $\overline{K}_x = \overline{k}_\mathrm{e} + \overline{k}_\mathrm{h}$. Note that optical excitation only allows exciton states with $K_x \approx 0$ because of momentum conservation. If screening is not large enough, Coulomb attraction will keep the two particles bound together in a hydrogenoid system. The two-particle state can be described in real space by the positions of its constituents, $\overline{r}_\mathrm{e}$ and $\overline{r}_\mathrm{h}$, with relative distance $\overline{r} = \overline{r}_\mathrm{e} - \overline{r}_\mathrm{h}$ and by the center of mass coordinate $\overline{R}_X = \frac{m_\mathrm{e}^* \overline{r}_\mathrm{e} + m_\mathrm{h}^* \overline{r}_\mathrm{h}}{m_\mathrm{e}^* + m_\mathrm{h}^*}$ (Figure 4.6). The total energy of the system is $E_{K_X,n} = E_\mathrm{g} - \frac{R}{n^2} + \frac{\hbar^2 K_X^2}{\left(m_\mathrm{e}^* + m_\mathrm{h}^*\right)}$ where

- E_g is the minimum energy required to generate free electron and hole (i.e., the semiconductor band gap);
- $-\frac{R}{n^2}$ (with $n = 1, 2, \ldots$) is the n-eigenvalue energy of the bound e–h pair, solution of the hydrogen-model Schrödinger equation with potential energy $U(\overline{r}) = \frac{1}{4\pi\varepsilon}\frac{e^2}{r}$. Bound states are located below the conduction band minimum, essentially lowering the optical gap of the material. R is the modified Rydberg constant $R = R_0 \frac{m_0}{\mu}\frac{1}{\varepsilon^2}$, with $R_0 = 13.6$ eV, ε dielectric constant of the semiconductor, and $\mu^{-1} = \frac{1}{m_\mathrm{e}^*} + \frac{1}{m_\mathrm{h}^*}$ the reduced mass of the system. The energy spectrum has a series of discrete states for increasing n till the merging into the conduction band continuum. The (Bohr) radius of the hydrogenoid, electron–positron atom in the semiconductor medium is $a_\mathrm{B} = a_\mathrm{B}^0 \varepsilon \left(\frac{m_0}{\mu}\right)$. Using $a_\mathrm{B}^0 = 0.5$ nm, $\varepsilon = 20, \mu = 0.1\, m_0\ a_\mathrm{B} \approx 5$ nm $\gg a$, where a is the lattice constant. The binding energy of the exciton is defined by $E_\mathrm{b} = E_1 - E_\mathrm{g}$;
- $\frac{\hbar^2 K_X^2}{\left(m_\mathrm{e}^* + m_\mathrm{h}^*\right)}$ is the center of mass translational kinetic energy within effective mass approximation. The parabolic dispersion law describes the free motion of the two-particle system in the covalent crystal.

The wavefunction associated with the state of energy $E_{K_X,n}$ can be expressed as

$$\Psi_X^{K_X,n}(\overline{R}, \overline{r}) = \frac{1}{\sqrt{V}} \mathrm{e}^{i\overline{K}_X \cdot \overline{R}} F_n(\overline{r}) \tag{4.11}$$

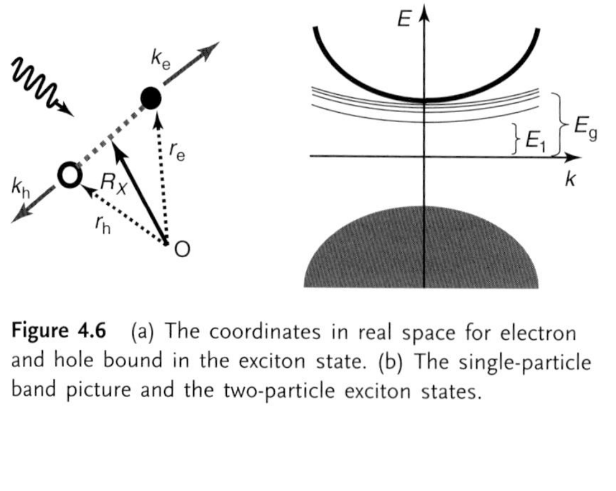

Figure 4.6 (a) The coordinates in real space for electron and hole bound in the exciton state. (b) The single-particle band picture and the two-particle exciton states.

The center of mass motion is described by a plane wave, corresponding to full delocalization in the crystal. When self-trapping takes place, this delocalization is lost and the center of mass of the exciton gets pinned to a trapping site of lower energy, a defect, or an impurity. The functions F_n describe the internal dynamics of the exciton, that is, the relative motion of the electron–hole pair. They are solutions of the hydrogen atom Schrödinger equation, with analogous series of s, p, and so on states (n represents the full set of quantum numbers n, l, m of the hydrogen atom solution).

Real excitations are coherent superpositions of excitonic states in the k-space, on a range Δk, which corresponds to wavepackets confined on a space $\Delta X = \frac{2\pi}{\Delta k}$. Figure 4.7 represents a simple picture for the exciton wavepacket. For optical excitation, when $\overline{K}_X = \overline{k}_e + \overline{k}_h \cong 0$, the spread in real space of the exciton wavepacket is $\Delta X \sim \lambda$ the wavelength of exciting radiation. The corresponding interval Δk contains many e–h pairs (all formed by vertical transitions and with near null total wavevector). For nonoptical excitation, for instance, formation of excitons on e–h recombination, $K_X \neq 0$ and the general rule is $\overline{K}_X = \overline{k}_e + \overline{k}_h = \left(\overline{k} + \frac{1}{2}\overline{K}_X\right) + \left(\frac{1}{2}\overline{K}_X - \overline{k}\right)$.

For optical excitation, $K_x \approx 0$, and the exciton state can be written as a linear superposition of conduction band–valence band (electron–hole) states according to $\Psi_X = \sum_k A(k) c_e(k) v_h(-k)$.

This reminds the same approach we followed in building the Frenkel exciton wavefunction if we consider that the electron–hole wavefunction is obtained from the ground-state wavefunction ($\Psi_0 = A\left\{v(k_1)v(k_2)\ldots v(k_n)\ldots v(k_N)\right\}$) switching one electron from the valence band to the conduction band, $\Psi c_{k_n}, v_{k_n} = A\left\{v(k_1)v(k_2)\ldots c(k_n)\ldots v(k_N)\right\}$. $A(k)$ weights the contribution of each k-state, and its Fourier transform is $F(r)$, the relative motion wavefunction.

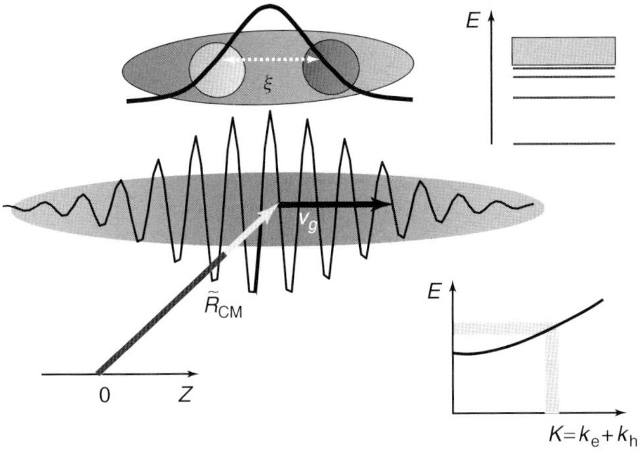

Figure 4.7 Wannier–Mott exciton. The center of mass wavepacket and the e–h relative motion are represented.

All e–h pairs that preserve the total exciton wavevector will contribute to the exciton state, filling up a corresponding volume in the k-space, known as *exciton volume*. The exciton as a quasiparticle is a boson, with total spin $S = 0$. In principle, excitons follow Bose–Einstein statistics and can give rise to phenomena such as Bose condensation. However, they are composite bosons made of a pair of fermions. Fermions follow Fermi–Dirac statistics and are subject to the Pauli blocking rule. As a consequence, the underlying fermionic nature may at some point emerge in the physical behavior of an exciton population. An example is phase space filling on optical exciton generation. Generating excitons in a material is similar to stuffing a box with hard balls. There is a finite number of excitons a system can support, determined by the occupation volume of the exciton. The exact calculation of the occupied volume should be done in the k-space where fractions of e–h states are "used" for building the exciton. In this calculation it is the fermionic nature of the electron and hole states that matters. It turns out that, a part for a numerical correction in the order of one, the volume in real space is the reciprocal of the volume in k-space. The exciton volume of a W–M exciton is the volume occupied by the e–h pair, approximately $\frac{4}{3}\pi\alpha_B^3$ in 3D, or $2a_B$ in 1D. A simple experiment to measure the exciton size is described below for the 1D exciton case.

The effect of excitons on the optical properties of the semiconductor is dramatic below gap, in the transmission region. We assume a two-band model for the semiconductor and consider the imaginary part of the dielectric function ε_2, which is proportional to absorption (Figure 4.8). Without excitonic effect we have

$$\begin{aligned} \varepsilon_2^F(\omega) &= 0 \qquad \hbar\omega < E_g \\ \varepsilon_2^F(\omega) &= \frac{C}{\omega^2}\left(\hbar\omega - E_g\right)^{\frac{1}{2}} \qquad \hbar\omega > E_g \end{aligned} \tag{4.12}$$

With excitonic effect this becomes

$$\begin{aligned} \varepsilon_2^X(\omega) &= \hbar\omega_n = E_g - \frac{R}{n^2} \qquad \hbar\omega < E_g \\ \varepsilon_2^X(\omega) &= \varepsilon_2^F(\omega)\,h(x) \qquad \hbar\omega > E_g \end{aligned} \tag{4.13}$$

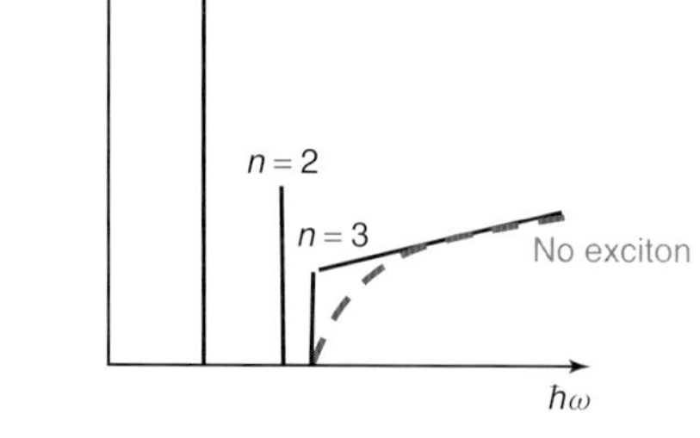

Figure 4.8 The exciton effect on band-to-band absorption.

where $h(x) = h\left(\sqrt{\frac{R}{\hbar\omega - E_g}}\right)$ is the above gap correction. The first equation describes a discrete spectrum with a series of lines, with decreasing oscillator strength $f_n \sim n^{-3}$ on increasing of $n = 1, 2, 3, \ldots$

Assuming constant dipole moment for the valence to conduction band transitions, M_{cv}, the exciton dipole moment square is $M_X^2 \propto |F_n(0)|^2 M_{cv}^2$, that is, it depends on the probability of finding the electron on top of the hole in the hydrogenoid atom.

Real excitations are wavepackets of excitonic states described by the coherent superposition

$$\Psi = \sum_{K_X} L(K_X)\Psi_X^{K_X,n} \tag{4.14}$$

Such an excitonic wavepacket propagates coherently in the crystal at group velocity $v_X = \frac{1}{\hbar}\overline{\nabla}_n E(K_X)$. The function $L(K_X)$ describes the probability amplitude of each K_X within the ΔK range, which defines the exciton band. The coherent transport as discussed for Frenkel exciton is still valid for the center of mass of the W–M exciton, once proper mass and dispersion laws are applied. Because the W–M effective mass is 1 or 2 orders of magnitude smaller than Frenkel exciton mass, the speed, and diffusion constant equally scale to higher values.

4.2.3 1D Excitons

The problem of the hydrogen model in 1D has been studied in 1950. Clearly the Coulomb potential $V(x)$ goes to negative infinite at origin, and the problem is not defined, with singular lowest wavefunction (a δ-function infinite at $x = 0$) (Figure 4.9). The mathematical oddity can be solved by introducing a cutoff to the Coulomb well. This has a straight physical meaning: real systems are never 1D. We call 1D a system whose aspect ratio L/d, where L is length and d is the

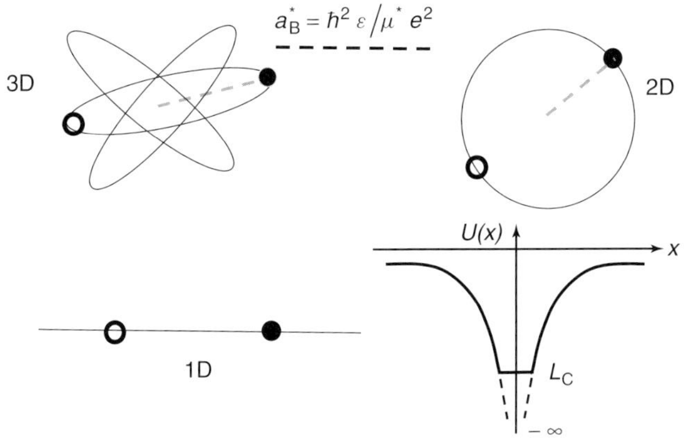

Figure 4.9 3D, 2D, and 1D hydrogenlike problem. Bottom right is the Coulomb well in 1D.

square root of a section, is very large. So 1D is an approximation. There will be a critical length, L_C, below which the 1D approximation fails, the aspect ratio approaches unity, and the point-charge approximation breaks down. Once the cutoff, L_C, is introduced, the problem can be analytically solved. The lowest exciton wavefunction is still reminiscent of the singularity at $x = 0$, and it is sharp peaked. The 1D wavefunction for $n = 1$ is

$$F_1(x) = \left(\frac{a}{L_C}\right)^{1/2} e^{-\frac{|x|a}{L_C}} \tag{4.15}$$

where a is the lattice constant. The dipole moment of the exciton transition is given by the sum of all the contributing e–h transitions in k-space, weighted by the exciton wavefunction $A(k)$, Fourier transform of $F(x)$.

$$|\mu_{Ex}|^2 = N\left|\sum_k A(k) M_{cv}(k)\right|^2 = NM_0^2\left|\sum_k A(k)\right|^2 = NM_0^2\,|F_1(0)|^2 \tag{4.16}$$

where we obtain the fundamental result that the dipole moment of the exciton transition is proportional to the e–h wave function at $x = 0$, that is, the probability that electron and hole are both in the origin. According to the exponential wavefunction F_1 we get

$$|\mu_{Ex}|^2 = \frac{a}{L_C} NM_0^2$$

For W–M exciton $L_C \gg a$, while for Frenkel exciton $L_C \approx a$, which implies that Frenkel excitons have much larger transition dipole moments.

The absorption spectrum of the 1D W–M exciton shows a very large intensity only in the first transition spectral line and an almost suppressed band-to-band transition (Figure 4.10).

Absorbance and saturation depends on the exciton size. In simplified terms, the absorbance A is proportional to the number of excitons the system can support, $A \sim N_X = Na/L_C$. This is called the *phase space filling model*. On photoexcitation, saturation sets in because the number of excitons that can be generated is

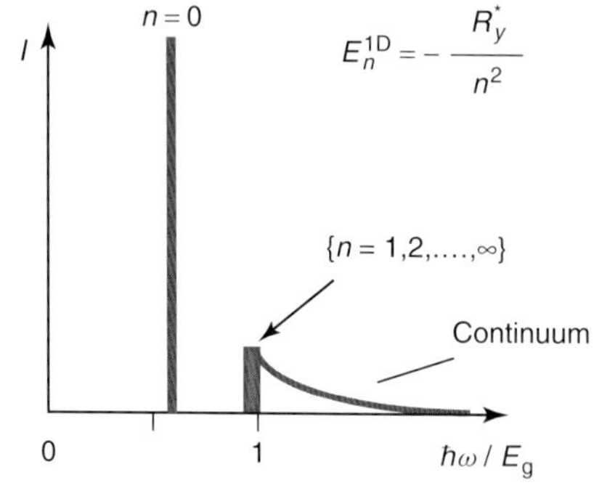

Figure 4.10 1D exciton absorption. The band-to-band transition above the gap is exaggerated, in real-life situation, it can disappear completely.

finite, $A^* \sim N_X(1 - \Delta N/N_X)$, where ΔN is the number of absorbed photons and generated excitons. The reduced absorbance under photoexcitation is called *photobleaching*, and it is experimentally accessible. This provides a way to measure experimentally the "size" of the exciton, or the critical parameter L_C. A simple expression is $\left(\frac{\Delta A}{A}\right)^{\exp} = -\frac{\Delta N}{N_X} = -nL_C$, where n is the linear density of photogenerated excitons. This model breaks down when exciton wavefunctions start to overlap and nonlinear interaction sets in. At a higher density, two (biexciton) or more (exciton strings) states may be squeezed into one.

Note that $F(0)$ is nonzero for s-like states, whereas p-like states will have zero amplitude at origin. This brings about a selection rule for one- and two-photon transitions, typical of excitons, which corresponds to a difference between the two-photon absorption gap and the one-photon absorption gap. In stark contrast, for the valence band to conduction band transition in semiconductors, the two-photon absorption gap coincides with the one-photon absorption gap.

Nonlinear optic depends, within the simple two-level model, on saturation. The different saturation behavior of Frenkel and W–M excitons reflects a different nonlinear response of the two states. The saturation density, N_X, is $N_X \approx N$ for Frenkel excitons and $N_X \ll N$ for W–M excitons. This matches perfectly with the relation for the oscillator strength as we stated above. It turns out that confined Frenkel excitons should have much large optical nonlinearities than extended W–M excitons.

5
Photoexcitation Dynamics

5.1
Photoexcitation and Relaxation Scenario in Inorganic Semiconductors

Photogenerated carriers in band states behave like a fermion gas. In stark contrast with molecules, on photoexcitation most of the energy is stored as kinetic energy of the carriers, whereas the phonon population remains initially unchanged. The carrier gas undergoes two types of relaxation: reequilibration by energy redistribution within the gas and relaxation by interaction with the lattice. Carrier relaxation phenomena are qualitatively similar in metal and semiconductors. Here we describe electron relaxation in the conduction band of a direct band gap semiconductor, with reference to Figure 5.1.

1) Photoexcitation by polarized light generates carriers within a narrow range of energy and in a narrow volume of *k*-space.
2) Momentum relaxation leads to loss of polarization memory (which is related to *k*) by redistributing *k*-vector into a random distribution. It takes place by elastic electron–electron scattering (there is no energy exchange, and no phonons are involved). It is an ultrafast process, typically occurring in about 10 fs.
3) Within 50 fs, thermal equilibrium among the excited carriers is established by electron–electron scattering (no phonon involved yet): electrons exchange energy during collision. The quasi mono energetic nascent distribution becomes a Boltzman equilibrium distribution at temperature T_c, much larger than lattice temperature, $T_c > T_L$. This distribution is called *hot carrier distribution.*
4) The carrier "plasma" loses energy by interaction with the lattice through electron–phonon coupling. This inelastic process leads to carrier energy loss and concomitant phonon emission. The phonon population builds up at a maximum rate $\left(h\nu_P^2\right)^{-1}$ where ν_P is the phonon frequency and $h\nu_P$ the energy emitted at each phonon cycle. Experimentally, if the excess energy to be dissipated for reaching the band bottom is ΔE, and the thermalization time of the electron gas is τ_e, one can estimate the average phonon emission time as $\Delta E/\tau_e$. This is usually of the order of the phonon period (e.g., 165 fs in GaAs), according to the simple picture proposed above. The build up and decay dynamics of the phonon population, $n(t)$, and its lifetime T_1^{vib}, can be

The Photophysics behind Photovoltaics and Photonics, First Edition. Guglielmo Lanzani.

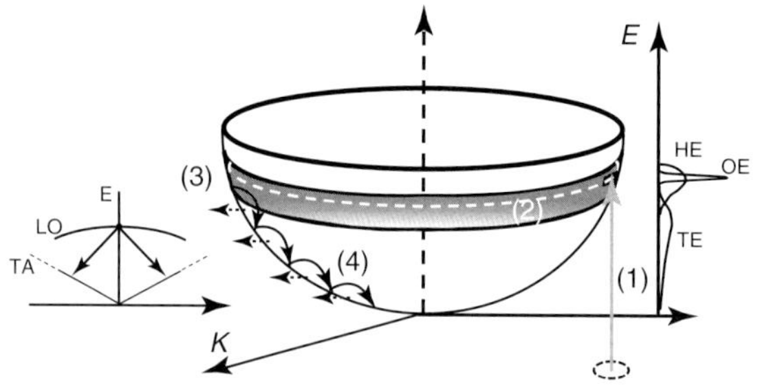

Figure 5.1 Electron relaxation in the conduction band. (1) Photon absorption by polarized light; (2) elastic electron–electron scattering leading to *k*-vector redistribution (10 fs); (3) inelastic electron–electron scattering leading to hot electron distribution (Te > TL) (100 fs); (4) electron–phonon scattering leading to electronic cooling and phonon population buildup (1 ps). Elastic (dephasing) and inelastic phonon–phonon scattering establish a hot phonon population in the lattice. Right axis: the initial electron out of equilibrium distribution (OE), the hot electron distribution (HE), and the thermalized state (TE). The inset on the left shows a simplified phonon dispersion plot, zoomed at the center zone, illustrating the Klemens decay model 1LO $\rightarrow$ 2TA.

measured directly by incoherent anti-Stokes Raman scattering. In this experiment, a pump pulse generates a hot carrier gas and a probe pulse measures the enhanced *spontaneous* anti-Stokes Raman scattering. A variant of the more common experiment of coherent anti-Stokes Raman scattering (CARS) that measures the *vibrational* coherence $Q(t)$ and the vibrational dephasing time (T_2^{vib}). (For a given Raman mode, $Q(t)$ is the collective nuclear displacement associated with the coherent Raman scattering signal $P^2_{CARS} \sim Q^2(t)$.) A drawback of the spontaneous anti-Stokes experiment is that Raman scattering can probe a very narrow range of phonon wavevectors, near $q = 0$, because of momentum conservation. At least in polar semiconductors however this is not a serious limitation because the electron–phonon coupling is due to Frölich interaction that depends on q^{-2}, where q is the phonon wavevector, generating phonons with prevailing small q. The photoinduced, spontaneous anti-Stokes Raman scattering has contributed important information about phonon dynamics, in particular on the so-called *bottleneck* effect. This is the slowing down of the carrier thermalization dynamics due to saturation of the phonon bath. The growth of the nonequilibrium phonon population is because of the longer lifetime of the phonon than the generation time. On population growth, the phonon emission rate slows down because of state filling, and so the dissipating efficiency of the phonon bath is reduced. Other relaxation mechanisms, such as Auger e–h scattering, are not efficient in bulk semiconductors, because of wavevector conservation. The situation is different in nanocrystals, where wavevector is not a good quantum number and momentum conservation is not an issue.

5) The carriers now dwell at the lowest energy (bottom of the band for electrons, top for holes), where further recombination or trapping may take place. Intervalley scattering is another process that can take place during or after the initial relaxation, displacing carrier population to another location in the *k*-space.

Optical phonons decay into acoustical phonons. An expression was derived by P. G. Klemens in 1966 [1] from perturbation theory for the lifetime of an optical phonon decaying into two acoustical phonons and its temperature dependence. Alternatively, a model was also proposed in which one longitudinal optical phonon decays into one transversal optical and one acoustical phonon. The general trend is, however, that energy flows into acoustical modes to contribute finally to increase the lattice temperature.

5.1.1 Link to Photovoltaic

Carrier relaxation is an important loss for power conversion. According to the Quessier–Shockley detailed balance analysis, each photon with above gap energy will contribute just the gap energy, losing all the excess as heat (phonons). This is partially responsible for the maximum conversion rate in the 30% range in spite of a much higher thermodynamic limit (85%). Hot carrier cells are those that extract carriers before thermalization, thus avoiding the dissipation of the excess energy with respect to the gap. One way to achieve this would be engineering the electron–phonon coupling in order to slow it down. Among several strategies, one invokes the bottleneck effect and transfers the concept to phonon relaxation. Hampering phonon decay would cause a buildup of the phonon population and accordingly a reduced efficiency in the dissipation of carrier energy. This is, however, quite a difficult task, which no one has still accomplished.

5.2 Confined States in Semiconducting Nanocrystals

In semiconducting crystals with size of the order of a few nanometers, electronic states "feel" the confinement and lead to a substantial modification of the electronic structure. In particular, below a critical size, characteristic of each material but in the order of 10 nm, the band structure rarefies into a set of discrete levels and the wavevector is no more a good quantum number. For excitonic semiconductors, the figure of merit is the nanocrystal size compared to the exciton Bohr radius.

The energy gap in a nanocrystal is enhanced with respect to the bulk value, by the characteristic confinement energy $E = E_g + \frac{\hbar^2\pi^2}{2\mu R^2}$, where R is the nanocrystal size and μ the reduced mass of the e–h pair (Wannier–Mott exciton). In addition, there is an extra contribution to the energy gap due to Coulomb interaction, $E_C \propto R^{-1}$, which is usually negative (bounding interaction) for a single-exciton state, but can be positive for biexciton or multiexciton states. Overall, the energy gap is $E = E_g + \frac{\hbar^2\pi^2}{2\mu R^2} + E_C$. Exciton–exciton interaction (both sign and strength), carrier relaxation, and carrier recombination are affected by quantum confinement. For instance, because carrier energy spacing is larger than typical phonon quanta, a substantial reduction in the dissipation efficiency of the phonon bath is expected. This has rarely been observed, because another effect comes into play, the fast

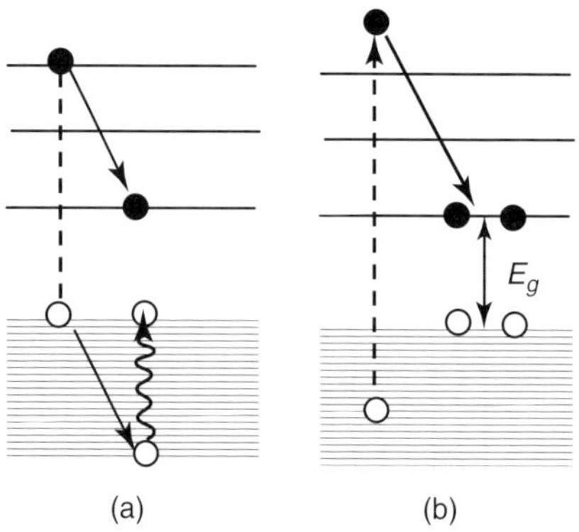

Figure 5.2 (a) Coulomb-interaction-mediated thermalization. The electron in some higher state gives away its energy to a hole, reaching the electron band bottom. The hole is initially scattered to a deep state (high energy), from where it relaxes back to the top state (lower energy) by efficient phonon emission. (b) Harvesting hot electrons by multicarrier generation. One hot e–h pair relaxes into two or more e–h pairs.

inelastic e–h Coulomb-interaction-mediated scattering. In nanocrystals this process is highly efficient because (i) e–h Coulomb interaction is stronger in small structures; (ii) the hole level spacing is much smaller than the electron one, making a dense manifold, almost a continuum, due to large hole effective mass and valence degeneracy; and (iii) holes quickly relax through the dense manifold of states emitting phonons (bulk mechanism). As a consequence, the process in Figure 5.2 takes place, leading to ultrafast relaxation of carriers to the frontier states.

Auger recombination is more efficient in nanocrystals than in bulk semiconductors because of the lack of momentum conservation as shown in Figure 5.3. The

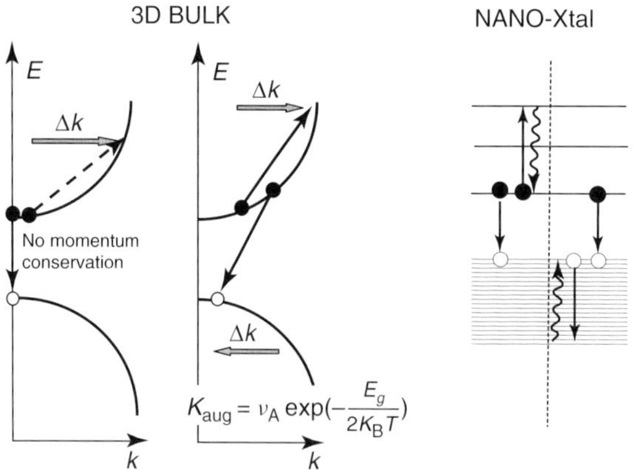

Figure 5.3 Auger recombination in the bulk is constrained by momentum conservation. It is possible at high temperature, when large k-vector states are populated; in QD there is no such limitation.

process, again Coulomb interaction mediated, consists of the interaction of an e–h pair with a third particle (either electron or hole). The pair recombines and gives away its energy to the third particle, which subsequently thermalizes. The rate of this process is thus dependent on the cube power of the carrier population. In the presence of many e–h pairs, this process is superlinear with power M, where M is the number of particles involved. The larger the number of particles the faster their recombination rate. This is the fundamental limitation to the application of carrier multiplication for photovoltaics as discussed below.

The electronic structure of nanocrystals depends on volume, shape, and composition. Typically used blends are based on II–IV semiconductors such as CdS or CdSe, but many others have been synthesized and studied. The shape of the nanoparticle can affect the electronic structure by introducing anisotropy in the quantum confinement. For instance, elongated structures such as rods are quasi 1D along the main axis and 0D across (Figure 5.4). In tetrapods or octapods, 0D and 1D confined states are mixed. The composition and the geometry contribute to define interfaces inside the nanocrystal. There are two well-defined situations, named type I and type II, as shown in Figure 5.5, and several intermediate cases. The level alignment at the interface is one of the most critical issues in defining the nanocrystal performance in possible applications.

Absorption of colloidal nanocrystals in liquid is described by a sum-over-state expression according to

$$\alpha(\omega) = N_{\mathrm{NC}} \times V_{\mathrm{NC}} \sum_{e} \sum_{h} a_{eh} \times g(\omega - \omega_{eh}) \tag{5.1}$$

where $a_{eh} = A_{eh}\omega_{eh} \, |\langle \phi_e | \, \phi_h \rangle|^2$contains the envelope overlap integral of the electron and hole wavefunction involved in the transition, A_{eh} is a constant containing the

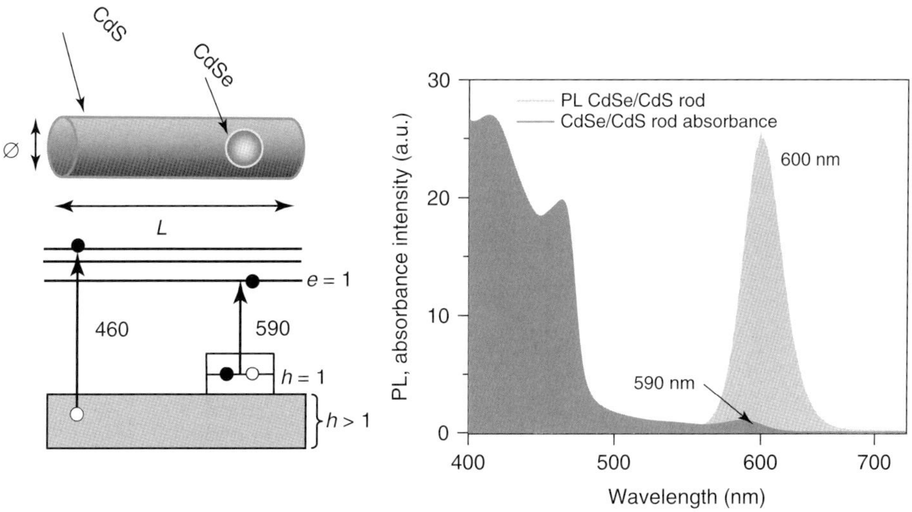

Figure 5.4 Optical transitions in CdS/CdSe quantum rods.

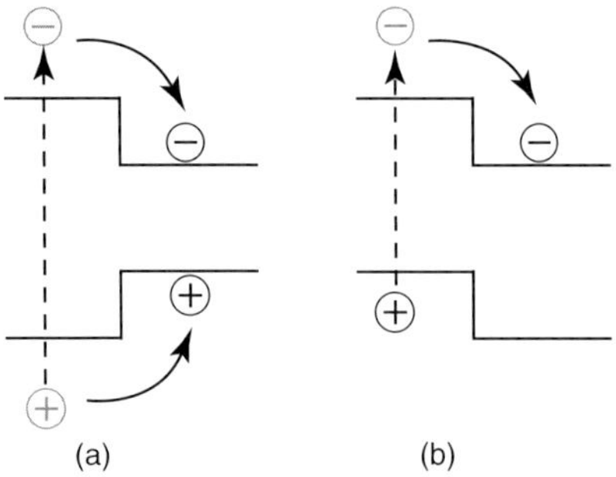

Figure 5.5 Type I (a) and type II (b) interface. Following photoexcitation (dashed arrow), type I leads to energy transfer, type II leads to charge separation (the symmetric process of hole transfer can be obviously designed).

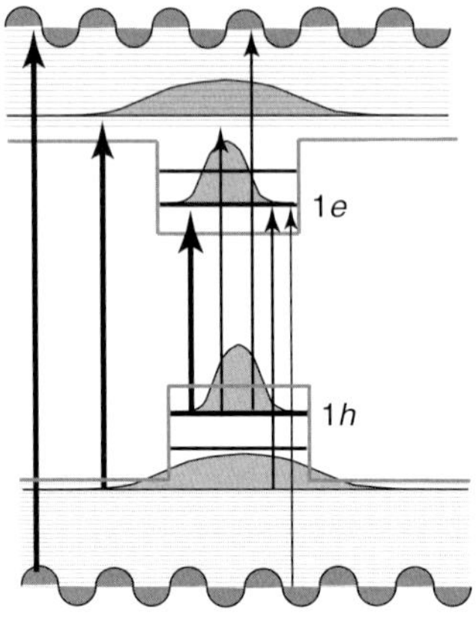

Figure 5.6 Schematic of wavefunction envelopes in confined states, with transition strength coded into arrow thickness. The picture suggests the role of the wavefunction envelope overlap in the transition oscillator strength for a rod-dot system with localized and delocalized states. Here, both electron and hole potentials show a clear band offset.

electronic dipole moment of the original atomic orbitals, N_{NC} is the nanocrystal concentration in solution, and V_{NC} is the nanocrystal volume. The overlap integral depends on the geometrical constrains on the initial and final states. As a result, localized states are weakly coupled to delocalized states because of different spatial extents. Figure 5.6 is an example of optical transitions in a dot-rod semiconductor nanocrystal.

On photoexcitation, there are three main contributions to the transient absorbance change: (i) state filling, (ii) Stark shift, and (iii) trapped induced transitions.

State filling is by far the main source of the transient signal. It is due to Pauli exclusion principle and the obvious outcome that empty states and filled states do not contribute to optical transitions. Following absorption, electrons (above gap) and holes (below gap) are formed. The absorbance difference is thus

$$\Delta\alpha(\omega) = -N_{\text{NC}} \times V_{\text{NC}} \sum_{e} \sum_{h} a_{eh} \times g(\omega - \omega_{eh}) \, (n_e + n_h) \tag{5.2}$$

where n_e is the electron occupation number ($n_e = 1/2$ for single occupancy) and n_h the hole occupation number. As an example, let us look at the simplified scheme in Figure 5.6. On pump excitation and eventually relaxation, the lowest e–h states are occupied ($e = 1$, $h = 1$). Bleaching will have several contributions that can be grouped like

$$\begin{aligned}\Delta\alpha(\omega) &= -A_{11}g(\omega - \omega_{11}) \, |\langle\phi_1| \, \phi_1\rangle|^2 + \frac{1}{2}\sum_{e>1} A_{e1}g(\omega - \omega_{e1}) \, |\langle\phi_e| \, \phi_1\rangle|^2 \\ &\quad + \frac{1}{2}\sum_{h>1} A_{1h}g(\omega - \omega_{1h}) \, |\langle\phi_1| \, \phi_h\rangle|^2 \end{aligned} \tag{5.3}$$

The first term is the transition between the highest hole state and the lowest electron state, with a large envelope function overlap; the second term contains all the transitions from $h = 1$ up, excluding the $e = 1$ state. On increasing the transition energy the function overlap decreases. The third term contains the transitions to $e = 1$ from all h-states, excluding $h = 1$.

- Stark shift of the transition energy can be induced by Coulomb interactions. In the presence of photoexcited carriers local electric fields can be generated, which affect the state energy, similar to the molecular Stark effect, or biexcitonic effects can be introduced. The latter again will appear as a shift in the transition energy. Both bound and repulsive states may be formed in nanocrystals because of different carrier localization at interfaces (for instance, core–shell effects). Very often Stark shifts are to the red, and photoinduced absorption (PA) appears below the optical threshold. In terms of biexciton it means that generating a second exciton costs less energy than for the first one. It happens quite often that on carrier relaxation the Stark-shift-induced PA is covered by the bleaching due to state filling of the frontier states.
- PA is due to trapped carriers (most of them on the surface or at the interface between two components). Their filling and depletion occurs on a multiscale dynamics. The fill-up usually happens during carrier thermalization or relaxation, on the subpicosecond or picosecond time scale. Their deactivation may span many decades in time, with lifetime extending to the millisecond time domain. Surface states do certainly play an important role in carrier dynamics, because in nano-objects the surface to volume ratio is much larger than in the bulk. To have an idea, just consider a cube of side 1 m. The surface is 6 m^2. Every time the side halved, the number of cubes is increased by a factor of 8 and the total surface doubles. If we keep cutting down the cubes till the side is 10 nm, the total surface of all the little cubes is 6×10^8 $m^2 = 600$ km^2. This is about the area of Corfu, a beautiful Greek island.

5.3
Carrier Multiplication in Nanocrystals

Carrier multiplication is the generation of more than one e–h pair on absorption of a single photon (Figure 5.2b). It is sometimes called multiexciton generation (MEG). In general, the required photon energy should be much larger than the optical gap. Following Oleg Prezhdo's recent review on the subject, we describe below three different mechanisms that can lead to multicarrier generation.

1) **Impact ionization process.** Carriers are generated by incoherent Coulomb scattering in which a high-energy carrier relaxes to its ground state and excites valence electrons across the band gap, producing additional e–h pairs in numbers consistent with the energy conservation rule. Roughly, the initial photon energy breaks down into e–h carriers with about gap energy each. This process is the reverse of Auger recombination, and the two compete. At high excitation energy, however, theory predicts that impact ionization is more efficient than Auger recombination and electron–hole relaxation. The Auger rate is enhanced by confinement, in spherical quantum dot according to a R^{-6} law.
2) **Coherent mechanism.** Here a photon of energy greater than twice the band gap creates a coherent superposition of degenerate single and multiexcitons states. Immediately following photoexcitation, the electronic population oscillates between the two types of states coupled by the Coulomb interaction. Once electron–phonon relaxation sets in, it introduces a dephasing mechanism for the coherent state. Assuming that the multiexciton state couples more strongly to phonons than the single-exciton state, the energy decay is faster when the system is in the multiexciton state, favoring its stabilization. The latter assumption is reasonable, since the multiexciton state regards a larger electron density change and consequently larger lattice distortion than a single-exciton state. Note the difference with electron–phonon coupling renormalization in the molecular exciton. The latter, however, regards intramolecular phonons.
3) **The direct mechanism.** A single absorbed photon could generate a multiexciton state directly, with a rate described by a high-order perturbative process. This requires that the electronic transition from ground state to the many-exciton state has a nonvanishing oscillator strength, provided again by Coulomb interaction, causing state mixing. The multiexcited state subsequently dissociates into low-energy singly excited states by electron-phonon-coupling-induced dephasing. In all three mechanisms, dephasing induced by electron–phonon scattering plays a crucial role.

Carrier multiplication in photovoltaics could avoid wasting of photon energy exceeding the gap. Usually this energy is dissipated into phonons (heat) and does not contribute to the yield in electrical energy. Harvesting hot carrier energy would boost the efficiency of the photovoltaic cell. Once formed, however, ultrafast recombination of the multiexciton states should be avoided in order to obtain an advantage in photovoltaics. This is the main culprit of the method. While carrier multiplication has been seen in the picosecond time scale, ultrafast recombination to a relaxed single e–h pair has so far hampered any practical use of this phenomenon. Note the similarity of this subject with the hot carrier cell discussed previously.

5.4 The Excitation Zoo in Molecular Semiconductors

Molecular semiconductors are a broad class of materials, which encompasses small conjugated molecules and polymers. Both can form crystals, amorphous solids, or aggregates or be isolated in solution or matrix. Figure 5.7 contains a list of their elementary excitations. Here we provide a brief description of each one.

5.4.1 Free Carriers

These are typical of crystalline solids with well-defined periodicity and space order. They are associated with Bloch states and propagate in the crystal as wavepackets.

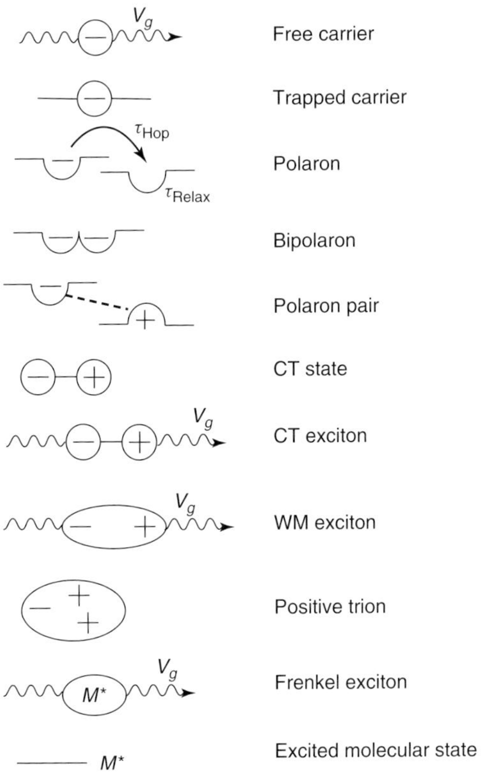

Figure 5.7 The elementary excitations of molecular semiconductors.

They carry a charge $\pm e$. They have a characteristic far infrared absorption, named Drude tail, which can be detected by using THz radiation, associated with transition within the quasi continuum of the conduction band. Free carriers are rarely observed in molecular semiconductors because of disorder, which induces localization.

5.4.2
Trapped Carriers

Defects in crystals such as structural defects (stacking faults, dislocations, and grain boundaries) or point defects (such as orientational, chemical, or impurity) are associated with localized levels, which may appear in the band gap of the semiconductor crystal. A carrier may relax into such lower energy sites and get trapped, losing its propagating character because the mobile Bloch-states wavepacket collapses into molecular-like wavefunction. Optical transitions can sometimes be associated with such localized charges. Mobility is affected by trap states in several ways: (i) they may cause a transition from coherent to incoherent transport, (ii) their energy distribution governs hopping, and (iii) their filling will determine the rate of trapping of the still free carriers (once filled traps are "inactive"). Temperature plays a crucial role in setting the effective trapping and detrapping rate. Shallow traps are easily freed by thermal activation, whereas deep traps are characterized by energy barriers larger than $k_B T$. Trap states may arise also at the surface, where dangling bonds are the major source of localization. As a consequence, interfaces are also keen to support such states.

5.4.3
Polarons

A polaron is a charge carrier associated with a local deformation of the lattice. This deformation can be due to electrostatic interaction in polar lattices or to electron-phonon coupling, as expected in conjugated polymers or one-dimensional solids in general. The polaron may preserve a delocalized nature and thus appear as a large-mass quasi particle or lose completely the wavelike nature and be in a fully localized state, still associated with the local geometrical relaxation. Self-trapping is the process of carrier-induced lattice deformation that leads to trapping of the carrier itself. It is not unique of charge excitations, but it can also occur with excitons. When the trapping time is faster than $h/\Delta E$, where ΔE is the bandwidth in energy, coherent transport is lost and hopping sets in. Polarons in conjugated oligomers are simply ions. With respect to the neutral molecule, an ion has a different chemical conformation. In this sense, a polaron is an ion in the amorphous solid, which propagates by hopping and carries a charge $\pm e$. Polarons carry spin $\frac{1}{2}$ (they are doublets). Usually removing one electron or adding one electron leads to redshift of the optical gap because of Coulomb interaction. New intragap electronic transitions that appear in the gap are fingerprint of the polaron (or other charged state) generation. In addition, when charges are added to the conjugated chain, some of the Raman active modes become infrared-active vibration (IRAV), since the symmetric oscillation may lead to an asymmetric charge displacement and thus to an electrical dipole that coupled to radiation. The IRAVs are manifested as peaks in the absorption spectrum; for this reason they are considered a fingerprint for charge excitations and used to discriminate those from neutral states. Both chemical doping or photoexcitation can lead to IRAV detection. The oscillator strength is huge because of the small kinetic mass of the localized charge that translates into a huge dipole. IRAVs are sharp peaks in energy. An interesting phenomenon appears when IRAVs are overlapped in energy to a broad electronic transition. Fano resonance between the two types of transitions leads to antiresonances, that is, dips in the broad absorption band.

5.4.4 Bipolaron

Two polarons of the same sign may coalesce and overcome the Coulomb repulsion, forming a weakly bound state named bipolaron. The attracting force counterbalancing the Coulomb repulsion is the lattice deformation and thus the extra gain in geometrical energy. A bipolaron can contribute to electrical current as polarons do, but most of the time it is localized. Bipolarons are associated with intragap electronic transitions, similar to polarons, but in principle with a different spectrum. In addition, they have different magnetic properties. Bipolarons are spinless excitations.

5.4.5 Polaron Pair

A distant pair of oppositely charged polarons may get bound because of Coulomb attraction. This bound pair, in which each unit preserves its nature (charge distribution and electronic and vibrational structure), is the polaron pair. Typically, it can take place between two adjacent conjugated chains carrying a charge each or between two segments of the same, coiled, polymer chain. The polar pair is a weakly bound state typical of amorphous aggregates. It has negligible wavefunction overlap between the two components (i.e., the hole–polaron and the electron–polaron). The two wavefunctions are virtually unaffected in the bound state. A polaron pair can have singlet or triplet spin multiplicity. Owing to the weak interaction these are quasi degenerate. A polar pair is geminate when it arises from neutral singlet ionization or non geminate as occurs on opposite charge encounter. In both situations, it is an intermediate state between full ionization and recombination. Polaron pairs have static dipoles, screened by the environment dielectric function, and weak or null optical coupling with ground state. Their absorption spectrum resembles that of an individual polaron, with a small shift in the transition energies. As a consequence it is difficult to distinguish an individual polaron from a polaron pair. Their kinetics of recombination is, however, different. A polaron pair population decays with monomolecular kinetics, a polaron population decays with bimolecular (density-dependent) kinetics.

5.4.6
Charge Transfer State

This is a molecular state with ionic character where regions of positive and negative charge are bound together by Coulomb attraction. The charge transfer (CT) between the two components is usually incomplete. CT is typical of a molecular aggregate or dimer, where neighboring molecules can interact. The CT is formed when a fraction of the charge is displaced, leading to a static dipole. The CT has its own identity different from that of the separated negative and positive ions. It can have spin singlet or triplet multiplicity. It has a wavefunction with both excitonic and ionic character. If the first largely prevails we will call it an exciton, if the second prevails we will call it a CT state, with all possible intermediate combinations. Large static dipoles are characteristic of CT states and may lead to a strong signal (second derivative in lineshape) in Stark spectra. CT states can be easily ionized and are considered as precursor of charge carriers. Donor–acceptor molecules form CT states upon photoexcitation. The electronic transitions of the CT state are different from that of the neutral molecule and also of the individual charge carriers. This marks the difference with polaron pairs. Because of the strong interaction, CT states have their own peculiar spectrum. CT states may have weak optical coupling with ground state and consequently show some radiative recombination. Their emission spectrum is often structureless.

5.4.7
CT Exciton

In molecular crystals, a CT state can have a wavelike nature and propagate coherently in the crystal as a quasiparticle. However, this remains a theoretical possibility, and most CT states are localized. The CT exciton is in between a Frenkel exciton and a Wannier exciton and pertains to molecular solids. A CT exciton has a strong static electrical dipole, easily spotted in Stark spectroscopy. CT excitons are neutral and currentless excitations, yet they can be easily ionized.

5.4.8
Wannier–Mott Exciton

This is a bound state of electron and hole in a band semiconductor. It is characterized by a fully delocalized center of mass whose wavelike propagation is described by wavepackets of Bloch periodic functions. The wavepacket amplitude square describes the probability to find the center of mass in the lattice. It also has an internal degree of freedom, associated with the relative asset of the electron and hole respect to their center of mass in the hydrogenlike system. A Wannier–Mott exciton, at variance with CT, does not have a static dipole. Wannier–Mott excitons are neutral, currentless excitations.

5.4.9 Trion (Charged Exciton)

In confined systems, where Coulomb interaction is enhanced, a second charge can be bound to the W–M exciton, giving rise to a charged particle. It can be formed by two positive charges and one negative charge or two negative charges and one positive charge. The trion will have a binding energy that shifts to the red its energy with respect to the neutral exciton. Observed in semiconductor heterostructures the trion is recently identified also in carbon nanotubes.

5.4.10 Frenkel Excitons

A Frenkel exciton is a collective excitation of a periodic arrangement of weakly interacting molecules. It is a wavelike excitation described by plane waves. Wavepackets coherently propagate in the crystal, according to the dispersion law. The amplitude squared of the wavepacket represents the probability to find the excited molecular state in the crystal. Frenkel excitons are neutral, currentless excitations.

5.4.11 Excited Molecular State

In amorphous polymer networks or fully disordered aggregates of molecules, each conjugated unit behaves like an individual molecule, which can be in a singlet or triplet excited state. Because they are in a densely packed aggregate, however, their energy is modulated giving rise to bands. In addition, such states can propagate, incoherently, by energy transfer from one molecule to another (Förster coupling for singlet, Dexter for triplet). For this reason, they are often called, improperly, *excitons*.

5.5 Conjugated Polymers

In conjugated polymers, out of many possible excitations as described above, there are three states that play a predominant role in photophysics, singlet, doublet, and triplet states (Figure 5.8). The singlet state in the table is the lowest above ground state, S_1. According to Kasha rule, this is the one from which emission can take place, and it is the one that collects population following relaxation. Other decay paths from higher lying singlets may lead to population of the triplet state (the lowest lying is T_1) and the doublet state, representing charge carriers (D_0). Bound states of doublets and their dynamics certainly play a crucial role

Name	Singlet S_1	Doublet D_0	Triplet T_1
Charge	0	$\pm e$	0
Spin	1	1/2	3
Lifetime	0.01–10 ns	10 ps–10 ms	100 ns–10 ms
Wavefunction	Delocalized	Delocalized	Confined (few monomers)
Deactivation	PL $S1 \rightarrow S0+h\nu$ ISC $S1 \rightarrow Tn$ IC $S1 \rightarrow S0$+heat	$D_1+D_1 \rightarrow S_0$+heat $D_1+D_1 \rightarrow S_1$+heat	PH $T1 \rightarrow S0+hn$ ISC $T1 \rightarrow S0$
Spectral signature	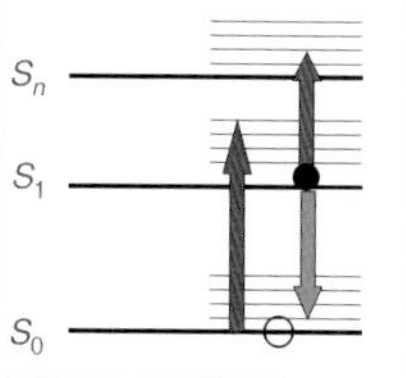	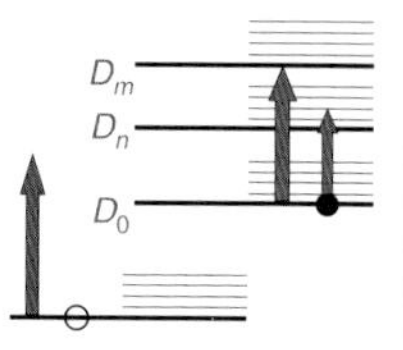	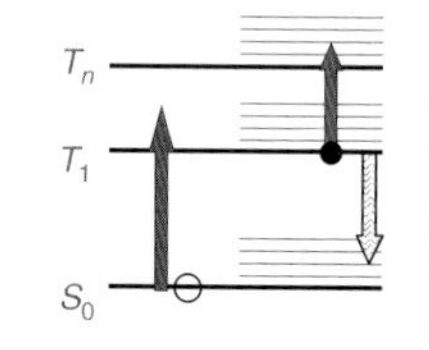

Figure 5.8 The three main elementary excitations in conjugated polymers.

in photovoltaics. One can see that many of the excitations mentioned above are essentially modifications of these three fundamental ones.

Singlet, doublet, and triplet states have wavefunctions delocalized over a certain number of molecular units. Triplets are more confined than singlet and polarons, with characteristic features in absorption that are narrow and with weak vibronic coupling. For all of them experiments show a size effect: the transition energy depends on the number of conjugated units in the linear chain, till some saturation. This observation is partially accounted for by the very simple particle in the box model. Accordingly, the energy structure in the rectangular 1D box is $E_n = \frac{h^2 n^2}{8mL^2}$, where L is the box size and m the mass particle (see any textbook of quantum mechanics for the solution of this problem). Assuming we have N carbon sites in the conjugated chain, each providing one π-electron, the first $n = N/2$ states are filled (allowing for spin). The energy gap is $\Delta E = \frac{h^2(n+1)^2}{8mL^2} - \frac{h^2 n^2}{8mL^2} = \frac{h^2}{8mL^2}(2n+1)$. Because the molecular size is $L = Na$, where a is the inter unit distance, the energy gap is $\Delta E = \frac{h^2}{8ma^2}\frac{N+1}{N^2} \approx \frac{B}{N}$ for $N \gg 1$. Plotting the transition energy ($S_0 - S_1$, $S_1 - S_n$, $T_1 - T_n$, $D_0 - D_m$) versus the inverse size ($1/N$), one finds that linear behavior is roughly fulfilled by many classes of conjugated systems (polyenes, carotenoids, thiophenes, phenylenes, phenyline–vynelines, etc.). The size parameter can be the number of double bounds or conjugated rings or other repeating units. On increasing the chain length, however, in all systems saturation sets in, and eventually the energy gap converges to a constant value. This is in contrast with $\Delta E \approx \frac{B}{N}$, which predicts a metallic state with $\Delta E = 0$ for an infinite long conjugated chain. The 1D metal is an issue in condensed matter physics, treated by Peierls, whose theorem states exactly that an infinite "metallic" chain with a half-filled band is unstable toward periodic distortion of the lattice pattern. The geometrical distortion opens up a semiconductor gap (Figure 5.9). The lowest order

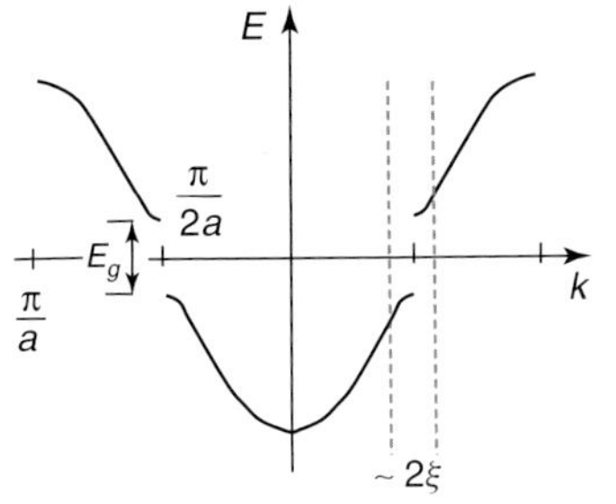

Figure 5.9 Peierls distortion opening a gap in the band structure of an ideal 1D system.

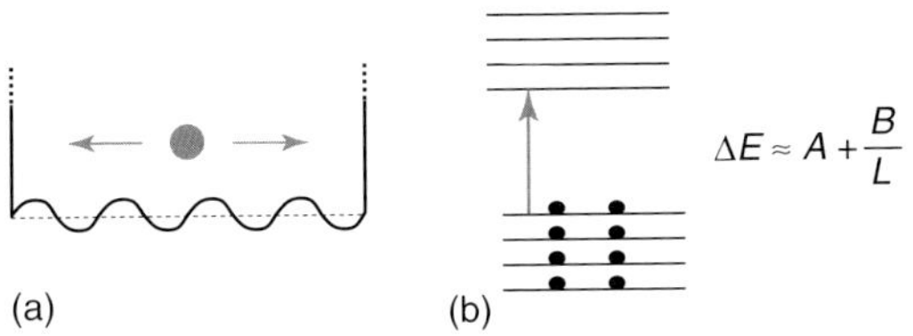

Figure 5.10 (a) The particle in the box model, with flat bottom (dashed) and corrugated bottom (Kuhn model, see text). (b) The electronic level structure.

perturbation to the 1D chain is dimerization, that is, pairing of sites in alternate single and double bonds. The very same results appear if one corrects the simple rectangular box potential with a cosine modulated potential energy and again solves the problem of the particle in the box (Figure 5.10). This is the famous Kuhn model, stated in 1949 [2], leading to the formula $\Delta E = A + \frac{B}{N}$, which correctly predicts a minimum energy gap, A, for infinite chain length. The tight binding Hamiltonian for the dimerized linear chain under cyclic boundary condition correctly predicts the semiconductor electronic structure. Two types of nearest neighbor overlap integrals, β_1 and β_2, are introduced for single and double bonds. The gap energy depends on the difference between the two overlap integrals, $E_g = 2(\beta_2 - \beta_1)$ and suggests a definition of the effective conjugation length as $L_C = \frac{\beta_2+\beta_1}{\beta_2-\beta_1}a$. On collapse of the dimerization, with $\beta_1 = \beta_2$, the gap would go to zero and the effective conjugation length to infinite, representing the fully delocalized, yet unstable, metallic state. To conclude, a physical alternation in the electron density of a very long conjugated chain leads, according to many different models, to a gap. A solid evidence for dimerization in conjugated chains comes from X-ray diffraction and Raman spectroscopy. Note here the difference with benzene, where the single–double structure is a theoretical artifact not found in experiment. In the 1D lattice theory, the effective conjugation length, L_C, specifies the segment in k-space, over which band bending takes place, causing the gap ($E_g \approx \frac{h\nu_F}{L_C}$, where ν_F is the Fermi velocity). The effective conjugation length is a fundamental parameter that specifies the excitation extension and corresponds to the exciton size in the 1D hydrogenoid atom problem (L_C is associated to the cutoff in the potential energy).

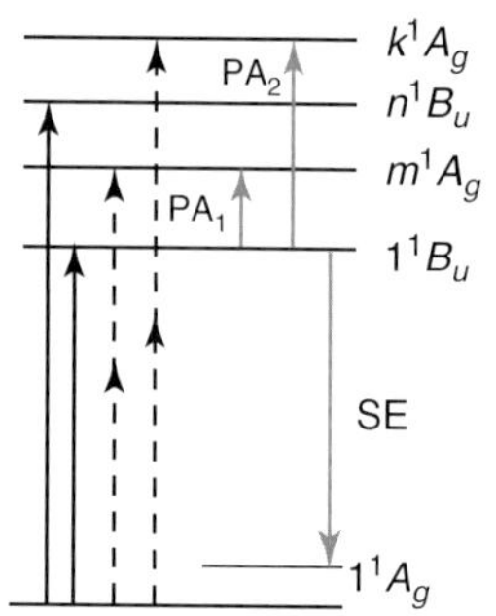

Figure 5.11 The five essential states in the singlet manifold for understanding many optical experiments in conjugated polymers. Ground-state transitions are black lines, representing one- and two-photon absorptions. Excited-state transitions are in gray, representing photoinduced absorption and stimulated emission. The level schemes regard an emitting polymer, with optically allowed lowest excited state. Vibrational levels are not shown. Doublet and triplet states are also omitted.

Therefore, from many different points of view, we see that a 1D system has a characteristic length that defines electron delocalization or exciton radius, below which the 1D approximation breaks down and 3D quantum confinement sets in. In molecular physics, the problem is most of the time discussed as "what happens when the chain size increases," while in solid state physics the typical problem is "what happens when the system size goes down." In a way, this is the dichotomy of the bottom-up or top-down approach in fabrication technology and reflects the natural occurrence of small molecules and bulk semiconductors. Because an infinite chain is anyway associated with a finite length, L_C, one can understand the blueshift of the optical gap on reducing the chain length as a quantum confinement effect. For instance, let us consider the thiophene ring. Going from the monomer to the dodecamer (12 rings), the absorption peak shifts to lower energy. For the longer chain the effects saturate, suggesting that 12 is about the size of L_C.

The real-world situation is more complicated than what we stated because the electronic motion in conjugated chains is controlled by three interactions: $\pi-\pi$ overlap leading to delocalization, electron–electron repulsion leading to localization, and electron–phonon coupling leading to carrier scattering. The role of electron correlation shows off dramatically in polyenes and carotenoids, where it is crucial for obtaining the correct energy level ordering. The lowest singlet dark state (S_1) can be properly computed only by advanced methods including double excitation and configuration interaction, that is, by accounting for correlation.

In conjugated polymers, a few excited states, named *essential*, can be highlighted in order to interpret a number of experimental results, including linear and nonlinear absorption, photoluminescence, and pump probe. Theory has been developed by Sumit Mazumdar, and many experiments validating it were done by Valy Vardeny. Figure 5.11 sums up their model, showing five states: ground state (S_0); first optically allowed excited state (S_1); S_n and S_k, both two-photon allowed; one-photon forbidden states coupled to S_1; and S_m optically allowed. According to C_{2h} symmetry, S_0, S_n, and S_k are A_g, while S_1 and S_m are B_u. Interestingly, this theory is fully tested by experiments:

1) peaks in the linear absorption can locate $1B_u$ and nB_u states (S_0-S_1), (S_0-S_2);
2) peaks in two-photon absorption spectrum will locate mA_g and kA_g (S_0-S_m), (S_0-S_k);
3) electroabsorption spectrum may contain features of all the sates;

4) PA bands in the pump probe spectrum are due to $1B_u - mA_g$ $(S_1 - S_m)$, and $1B_u - kA_g$ $(S_1 - S_k)$;
5) nonlinear photoconductivity, properly plotted, shows mA_g, nB_u, and kA_g.

Singlets, triplets, and doublets can interact among each other. Some bimolecular reactions that could occur are listed below. The list, self explanatory, is not exhaustive.

$$S_1 + S_1 \rightarrow S_n + S_0 \rightarrow S_1 + S_0$$
$$S_1 + S_1 \rightarrow S_n + S_0 \rightarrow D_0^+ + D_0^- + S_0$$
$$S_1 + S_1 \rightarrow S_n + S_0 \rightarrow T_1 + T_1 + S_0$$
$$T_1 + T_1 \rightarrow T_n + S_0 \rightarrow T_1 + S_0$$
$$T_1 + T_1 \rightarrow S_1 + S_0$$
$$S_1 + T_1 \rightarrow T_n + S_0 \rightarrow T_1 + S_0$$
$$S_1 + D_0^+ \rightarrow D_n^+ + S_0 \rightarrow D_0^+ + S_0$$

All these reactions are driven by energy transfer either by dipole–dipole coupling (Förster) or by electron exchange (Dexter).

References

1. Klemens, P.G. (1966) *Phys. Rev.*, **148**, 845–848. (A model for phonon decay).
2. Kuhn, H (1949) *J. Chem. Phys.*, **17**, 1198. (The electron in the box model).

6
Photophysics Tool Box

6.1
Jablonski Diagram

Once a molecule absorbs a photon, making a transition to a vibronic excited state, many processes can take place, eventually leading to reequilibration and excited-state deactivation back to the ground state. In Figure 6.1, the so-called Jablonski diagram is reported for unimolecular processes. Internal conversion (IC) is the nonradiative transition between states of same spin multiplicity, while intersystem crossing (ISC) regards transitions between different spin manifolds. Vibrational relaxation (VR) is the process of redistribution of vibrational energy from optically coupled modes to "dark" modes. These three processes do not involve the electromagnetic field. The transition occurs between equienergetic molecular states; thus it is represented by a horizontal arrow. Absorption, fluorescence, and phosphorescence involve coupling with photons. These transitions are represented by vertical lines.

The total rate of depopulation of an excited state due to p different relaxation processes is the sum of the rates of each process, $k_{\mathrm{T}} = \sum_{i=1}^{p} k_i$. Each process "$j$" has efficiency (or probability) $\eta_i = \frac{k_i}{k_{\mathrm{T}}}$. This simple equation states that the fastest "win," and it is a fundamental justification for time-resolved spectroscopy. Experience suggests the typical time scales for the most common molecular processes:

$$\begin{aligned}
k_{\mathrm{VR}} &\approx 10^{12}\text{–}10^{13}\ \mathrm{s}^{-1} \\
k_{\mathrm{IC}} &\approx 10^{12}\text{–}10^{14}\ \mathrm{s}^{-1}\ (S_n \rightarrow S_1) \\
k_{\mathrm{IC}} &\approx 10^{8}\text{–}10^{10}\ \mathrm{s}^{-1}\ (S_1 \rightarrow S_0) \\
k_{\mathrm{ISC}} &\approx 10^{6}\text{–}10^{11}\ \mathrm{s}^{-1}\ (S_1 \rightarrow T_1) \\
k_{\mathrm{ISC}} &\approx 10^{4}\text{–}10^{1}\ \mathrm{s}^{-1}\ (T_1 \rightarrow S_0)
\end{aligned}$$

The Jablonski diagram is a useful representation, yet an over simplification of what really happens in molecules. The main missing information is the coordinate-dependent state crossing, occurring in the multidimensional configurational space. Phenomena such as nonadiabatic crossing at conical intersections are missed; as a consequence, the predicted relaxation path can be wrong. In addition, the diagram does not contain coupling with the environment. Any real

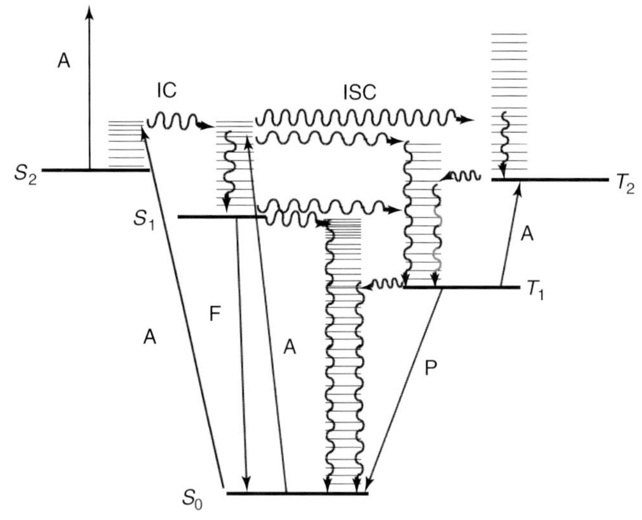

Figure 6.1 Jablonski diagram for unimolecular processes in the optically excited molecule. (i) IC: internal conversion, (ii) ISC: intersystem crossing, (iii) VR: vibrational relaxation, (iv) A: absorption, (v) F: fluorescence, and (vi) P: phosphorescence.

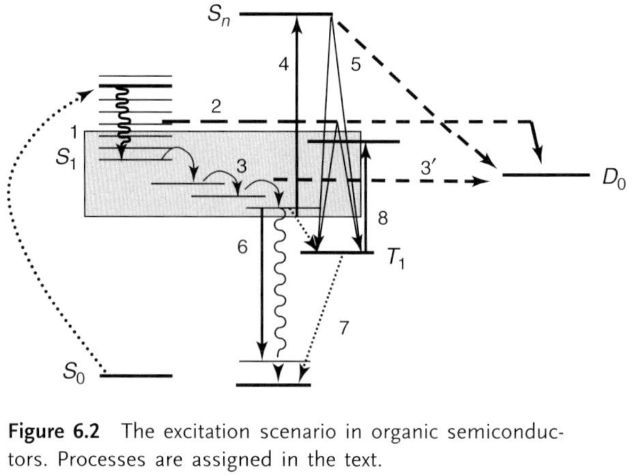

Figure 6.2 The excitation scenario in organic semiconductors. Processes are assigned in the text.

system is at thermodynamic contact with a bath that is responsible for transition energy renormalization and vibrational cooling (VC). VC is the dissipation of the molecule's internal energy to the environment.

Figure 6.2 is an extended Jablonski diagram, proposed for conjugated polymer semiconductors.

Process "1" is vibrational energy redistribution. It consists of down conversion of the vibrational energy, initially stored in few optically coupled modes, to a manifold of lower energy vibrational modes. Vibrational anharmonicity drives it. As a result, the electronic state will appear "relaxed" with respect to optically coupled modes. Note, however, that the initial excess energy at this time still dwells on the absorbing molecule, which is "hot," that is, its vibrational population distribution corresponds to a temperature higher than that of the environment. The time scale of this process is typically few hundred femtoseconds.

Process "2" is dissociation of the neutral state (autoionization) into charged pairs, which typically stay correlated, or into triplet pairs singlet fission (SF). Both may require excess vibrational energy with respect to relaxed singlets to fulfill energy conservation: final state energy $\leq$ initial state energy.

Process "3" is the intermolecular process of energy migration through the density of states. During this process, disorder-induced charge separation can take place, when neighboring molecules have energy levels aligned as in donor–acceptor (DA) pairs, because of local constraints. Energy migration takes place in about few picoseconds.

Process "4" is singlet–singlet annihilation. It leads to population of a higher lying singlet and thus a renewed deactivation process. Initially, energy will be exchanged with lower lying singlet states, a process called *internal conversion*, and a situation similar to "1" will occur.

Process "5" is fission from S_n, either autoionization into a charge pair or fission into triplet pair, from a higher lying excited state. It competes with internal conversion.

Process "6" is radiative and nonradiative decay of the lowest singlet. The former typically has a time constant of 1 ns and leads to fluorescence, and the latter can vary from nanoseconds to hundreds of picoseconds or even few picoseconds. As explained below, this determines the final outcome.

Process "7" is decay of the lowest triplet state. Triplet states may have lifetimes extending to the millisecond time scale. Phosphorescence is the weak radiative decay of triplets.

Process "8" is bimolecular recombination of triplets to generate a singlet state. It is an encounter reaction mediated by diffusion or the recombination of the triplet pair formed by fission.

6.2
Strickler–Berg Relationship

The Strickler–Berg (S–B) equation allows estimation of, under certain approximations, the radiative decay time of a molecular state from the calibrated *cw* absorption and emission spectra. In general terms, a molecular state decays through radiative (R) and nonradiative (NR) paths, with a total deactivation rate $K = \frac{1}{\tau} = K_R + K_{NR}$. For dipole allowed transitions K_R is in the order of 1–10 ns. $K_{NR} = \frac{1}{\tau_{NR}}$ is the sum of many processes, as mentioned above, that usually contribute to make the actual state lifetime much shorter than the radiative decay. The photoluminescence (PL) quantum yield η_F (number of emitted photons per absorbed photons) is given by $\eta_F = \frac{K_R}{K_R + K_{NR}}$. Obviously only for $K_{NR} = 0$ $\eta_F = 1$. Note that a unity quantum yield

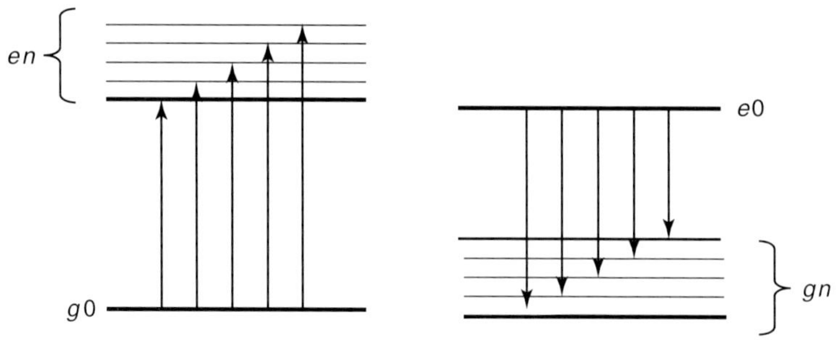

Figure 6.3 A molecule absorbs and reemits photons through the same vibronic manifold, according to g0→ *en* and *e*0→ *gm*, respectively.

is independent from the value of K_R, which can be very small (i.e., the transition can be weakly allowed and still have a unitarian emission quantum yield). It is the competition between radiative and nonradiative decay that dictates the overall emitting performance of the system.

In its extreme approximation, the S–B equation couples the transition oscillator strength to the radiative decay rate, $\frac{1}{\tau_R} = \frac{3}{2}\overline{\nu}_0^2 f$, where τ_R is in seconds, $\overline{\nu}_0$ is the peak wavenumber of the absorption transition, in cm^{-1}, and f is the oscillator strength of the absorption transition ($f = \frac{4.39\times10^{-9}}{n}\int \varepsilon(\overline{\nu})d\overline{\nu} = \frac{1.47\times10^{-19}}{n}\int \varepsilon(\nu)d\nu = \frac{3.85\times10^{-1}}{n}\int \sigma(\nu)d\nu$). Assuming an absorption peak at 500 nm, corresponding to 20 000 cm^{-1}, oscillator strength $f = 1$ (strongly dipole allowed molecular transition), and index of refraction $n = 1$, the S–B equation predicts $\tau_R = 1.6$ ns. For a forbidden transition, with $f = 10^{-6}$, the radiative decay time is 4 μs, as typically found for phosphorescence from the triplet state.

The S–B equation is obtained (as described in appendix 6.A) starting from the situation shown in Figure 6.3, where a molecule absorbs and reemits photons through the vibronic transitions g0 → *en* and *e*0→ *gm*, respectively. The full S–B equation, as reported in the original article of 1962 and without taking into account possible degeneracy of the molecular state, is

$$\frac{1}{\tau_R} = 2.88 \times 10^{-9}\frac{n_F^3}{n_A}\left\langle\overline{\nu}_F^{-3}\right\rangle^{-1}\int \varepsilon(\overline{\nu})d(\ln\overline{\nu}) \tag{6.1}$$

Here, n_F is the average refractive index over the emission band, n_A is the average refractive index over the absorption band, the expression in brackets is the inverse cube average wavenumber of the emission, weighted on the PL spectrum:

$$\left\langle\overline{\nu}_F^{-3}\right\rangle^{-1} = \frac{\int \phi(\overline{\nu})d\overline{\nu}}{\int \frac{\phi(\overline{\nu})}{\overline{\nu}^3}d\overline{\nu}} \tag{6.2}$$

The function ϕ is the fluorescence spectrum, in number of emitted photons per unit of frequency and absorbed photons, that is, ϕ is connected to the PL quantum yield η_F (number of emitted photons per absorbed photons) by $\eta_F = \int \phi(\nu)d\nu$.

The choice of the extinction coefficient in wavenumbers is typical for spectroscopy of molecules in solution, which is indeed the best situation for using S–B. The relationship can be stated in many different ways, for instance, using the numerical relationship between the extinction coefficient ε (cm^{-1} mol^{-1} l) and cross section $\sigma(\text{cm}^2)$, $\sigma = \frac{2303}{N_A}\varepsilon$ one can write

$$\frac{1}{\tau_R} = \frac{8\pi n^2}{c^2}\nu_0^3 \int \frac{\sigma(\nu)}{\nu}\text{d}\nu \cong \frac{8\pi n^2}{c^2}\nu_0^2 \int \sigma(\nu)\text{d}\nu$$
$$= 2.79 \times 10^{-20}\nu_0^2 \int \sigma(\nu)\text{d}\nu \tag{6.3}$$

where ν_0 is the average emission/absorption frequency and we assume narrow band ($\frac{\Delta\nu}{\nu_0} \ll 1$) for the first approximation and $n = 1$ for numerical evaluation.

The S–B relationship is obtained under one important assumption: mirror symmetry. The operative definition states that there is mirror symmetry between absorption and emission when the absorption spectrum reversed on the frequency axis and normalized by the cube of the frequency overlaps the emission spectrum. Physically this occurs when the same manifold of vibronic states is the *origin* of emission and the *destination* of absorption, as depicted in Figure 6.3. In quantitative terms, Strickler and Berg assume that the following relationship holds true for the B Einstein coefficients: $B_{g0\to en} = B_{e0\to gm}$, where g and e are the ground and excited electronic states, respectively, and n or m the vibrational quantum numbers $(0, 1, 2, \ldots)$. The Einstein coefficient B enters in the definition of the stimulated transition rate (number of transitions per unit of time) for a vibronic transition $W_{la\to ub} = \frac{\text{d}N}{\text{d}t} = N_{la}B_{la\to ub}\rho(\nu_{la\to ub})$, and it is expressed in m^3 J^{-1} s^{-2} if $\nu_{la\to ub}$ is the frequency of the vibronic transition (s^{-1}), $\rho\text{d}\nu$ is the radiation energy density (J/m^3), and N_{la} is the number of molecules in state la per unit volume (m^{-3}). B is a measure of the coupling with light, as clear from the relationship $B_{lu} = K\,|\mu_{lu}|^2$ where μ_{lu} is the mean electronic transition moment and K a constant. B is responsible for absorption or stimulated emission. Einstein also considers that a molecule in state ub may undergo a spontaneous decay to la, see Figure 6.4, so that his expression for the overall downward transition rate is

$$W_{ub\to la} = \frac{\text{d}N}{\text{d}t} = N_{ub}\left[B_{ub\to la}\rho(\nu_{ub\to la}) + A_{ub\to la}\right]$$

where A is the spontaneous transition rate, $A = 1/\tau_R$.

At thermodynamical equilibrium the transition rates from and to ub are equal (see appendix 6.A), leading to

$$A_{ub\to la} = 8\pi h N_{ub}\nu^3_{ub\to la}n^3c^{-3}B_{ub\to la} \tag{6.4}$$

The cube frequency term, which is present in the radiation energy density, is consistent with the characteristic ν^4 dependence of the power emission rate of the oscillating dipole (the emitted power is $\propto h\nu \times A$). Equation (6.4) is a very important and fundamental relationship, telling us that a system that is strongly coupled to light will also have a high rate of spontaneous decay, that is, a short radiative decay time.

According to the original Einstein treatment of 1917, the stimulated transition rates upward and downward are equal, as a consequence of the very general

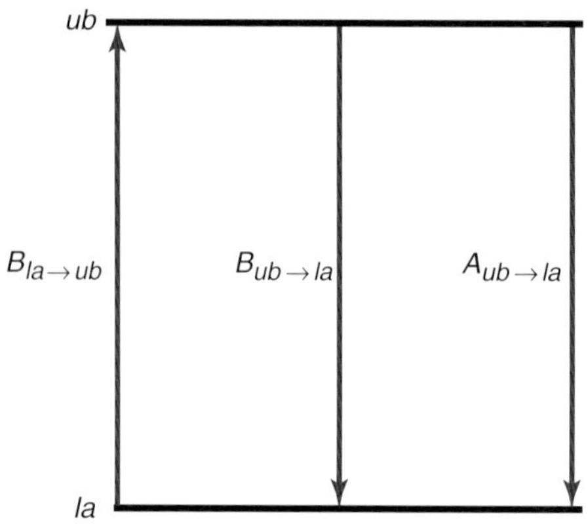

Figure 6.4 The *A* and *B* Einstein coefficients.

invariance of the two situations, or $B_{la\rightarrow ub} = B_{ub\rightarrow la}$, where *la* and *ub* are generic vibronic states.

This is a fundamental relationship, with general validity, while mirror symmetry is a particular case. Note the difference between mirror symmetry ($B_{g0\rightarrow en} = B_{e0\rightarrow gm}$) and Einstein equality ($B_{la\rightarrow ub} = B_{ub\rightarrow la}$). Mirror symmetry between emission and absorption spectra will occur if the emitting vibronic state is an exact replica of the ground state, simply shifted in energy and eventually displaced in configurational space. In real case, there is always a Stoke shift between absorption and emission gap, that is, an energy difference between the 0–0 lines, which accounts for a few percentage of the transition energy. Stokes shift is typically caused by readjustment of the solvent cage in the excited state, leading to energy renormalization.

The extreme approximation mentioned at the very beginning of this paragraph, leading to $\frac{1}{\tau_R} = \frac{3}{2}\bar{\nu}_0^2 f$, consists in assuming that absorption and emission spectra are very narrow and with negligible Stokes shift. Accordingly, one can use a single number, ν_0, for both the emission and absorption frequencies. This reduces the integral in the S–B equation to the oscillator strength definition. By further assuming refractive index $n = 1$, one obtains the short expression.

In practice, once absorption is measured, one can estimate the radiative decay time using S–B. On measuring the PL decay time ($K = 1/\tau$) it is possible to estimate the emission quantum yield. The advantage of this procedure is to avoid the direct experimental measurement of the emission quantum yield, which requires the use of an integrating sphere for collecting the emitted radiation together with a standard for calibration. The S–B equation allows discriminating between the mirror symmetry situation (when S–B works) and the emission from a different electronic state, possibly after major relaxation phenomena (when S–B does not work). Examples of the latter situation are found in weakly or nonemitting linear conjugated chains, such as carotenoids, polyenes, and some semiconducting polymers (notably those with topologically degenerate ground state). In these systems, the lowest excited state is not responsible for absorption due to lack of

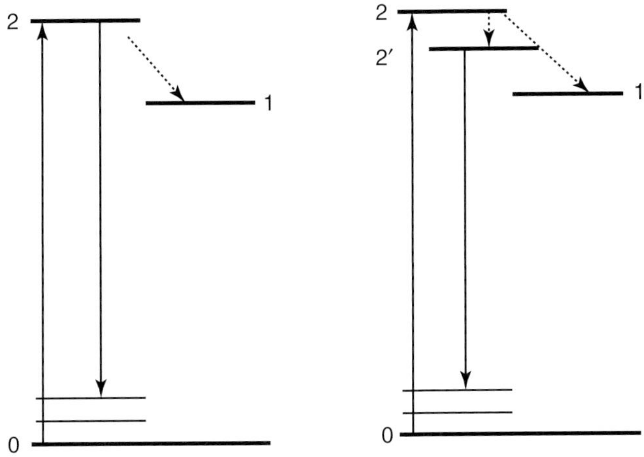

Figure 6.5 Two possible relaxation scenarios. (a) State 2 decays nonradiatively to dark state 1 and radiatively to 0. From S–B radiative time is 4.5 ns; 2–1 occurs with time constant 70 fs, and expected quantum yield is 10^{-5}. (b) The decay of state 2 is still 70 fs, but now there is a new path 2–2′ with time constant 7 ps. If 2′–0 is only radiative, the quantum yield of emission is 10^{-2}. If the energy gap between 2 and 2′ is not large, the emission spectrum shift might be exchanged for an extrinsic Stoke shift due to solvent rearrangement.

dipole coupling. The first optically allowed state is above ($S_{n\geq 2}$). Following, the absorption energy relaxes to the lowest, dark state (S_1). Emission may still occur from the higher optical state (anti-Kasha emission) for longer chains or from the vibronically activated S_1 state (Herzberg–Teller coupling) for short chains, respectively. Using S–B, one can estimate from absorption (S_0–S_2) the radiative decay time $K_R(S_2)$ of the optically active state (for instance, assuming emission at the same energy of absorption). From this and the measured decay time of emission ($K(S_2)$) it is possible to estimate the nonradiative decay responsible for population of the dark state using $K = K_R + K_{NR}$, without embarking into the difficult direct measurement of η_F.

In general, using the S–B equation helps in understanding the relaxation paths and the underlying electronic level structure. An example is the case of substituted polydyacetylene as simplified in Figure 6.5. Pump probe experiments indicate that the buildup time of photoinduced absorption from state $|1\rangle$ is about 70 fs. $1\rangle$ is a dark state (triplet pair) with negligible radiative coupling with ground state. The S–B relationship for state $|2\rangle$ yields $\tau_R = 4.5$ ns, and consequently the estimated fluorescence yield (from state 2 to ground state) is $\eta_F \approx 10^{-5}$. The experimental η_F is, however, about 1%, that is, three orders of magnitude larger. This suggests that the three-state model is not correct. A different model is based on four levels. After excitation, ultrafast branching leads to population of state $|1\rangle$ and $|2'\rangle$. If buildup time is 70 fs and emission yield is 1%, then the relaxation $|2\rangle$-$|2'\rangle$ occurs with time constant $\tau_{|2\rangle-|2'\rangle} \leq 7$ ps, with the longest value for unity quantum efficiency from $|2'\rangle$.

6.3
Energy Migration and Transfer

Excitonic wavelike propagation in crystalline solids is not the only way of energy propagation in a molecular ensemble. Many processes can lead to energy propagation, in which the coherent nature of the transport process is lost and energy "jumps" from one site to another.

According to J.B. Birks, energy transport between molecules can be differentiated as energy transfer, occurring between molecules of different species, and energy migration, occurring between molecules of the same species.

Energy migration or transfer can occur by *radiative process*, involving the emission of a photon by the donor molecule and its subsequent absorption by the acceptor molecule, often called *reabsorption*. It may also occur by a *radiationless process* due to interaction between the donor and the acceptor molecule during the excited-state lifetime of the donor, before its emission of a photon. Radiationless migration or transfer due to Coulomb (mostly dipole–dipole) interaction may take place over intermolecular distances (~1–10 nm) that are large compared to the molecular size. Radiationless migration or transfer due to electron-exchange interaction may take place over intermolecular distances (0.5–1.5 nm) that are somewhat larger than the molecular diameter. The latter, however, requires wavefunction overlap to occur efficiently. In fluid solution or at interfaces, collisional interaction due to Coulombic, electron-exchange, exciton resonance, and charge transfer interactions may yield excimers or exciplexes, the dissociation of which provides a further mechanism of radiationless energy transport.

The principal mechanisms of energy migration or transfer from a donor molecule D to an acceptor molecule A can be described as follows: (If D and A are of the same species, we are describing migration, if they are of different species, transfer. When necessary we will directly specify the molecules involved.)

1) Radiative transfer or migration

$$^{1}\mathrm{D}^{*} \longrightarrow {}^{1}\mathrm{D} + h\nu_{\mathrm{D}};\ {}^{1}\mathrm{A} + h\nu_{\mathrm{D}} \longrightarrow {}^{1}\mathrm{A}^{*}$$

Triplet–triplet and singlet–triplet radiative migration are usually negligible because of the low absorption intensity of S_0-T_1. Triplet–singlet radiative transfer may, however, take place:

$$^{3}\mathrm{D}^{*} \longrightarrow {}^{1}\mathrm{D} + h\nu_{\mathrm{D}};\ {}^{1}\mathrm{A} + h\nu_{\mathrm{D}} \longrightarrow {}^{1}\mathrm{A}^{*};$$

2) Collisional migration due to excimer formation and dissociation (A and B are molecules of the same species)

a. **Singlet–singlet:**

$$^{1}\mathrm{M}_{\mathrm{A}}^{*} + {}^{1}\mathrm{M}_{\mathrm{B}} \longleftrightarrow {}^{1}\mathrm{Ex}_{\mathrm{AB}}^{*} \longrightarrow {}^{1}\mathrm{M}_{\mathrm{A}} + {}^{1}\mathrm{M}_{\mathrm{B}}^{*}$$

b. **Triplet–triplet**

$$^{3}\mathrm{M}_{\mathrm{A}}^{*} + {}^{1}\mathrm{M}_{\mathrm{B}} \longleftrightarrow {}^{3}\mathrm{Ex}_{\mathrm{AB}}^{*} \longrightarrow {}^{1}\mathrm{M}_{\mathrm{A}} + {}^{3}\mathrm{M}_{\mathrm{B}}^{*}$$

3) Collisional transfer due to exciplex formation and dissociation (M and Y are different molecular species)
   a. **Singlet–singlet:**

$$^1M^* + {}^1Y \longleftrightarrow {}^1(M \cdot Y)^* \longrightarrow {}^1M + {}^1Y^*$$

   b. **Triplet–triplet:**

$$^3M^* + {}^1Y \longleftrightarrow {}^3(MY)^* \longrightarrow {}^1M + {}^3Y^*$$

4) Radiationless migration or transfer due to electron exchange
   a. **Singlet–singlet:**

$$^1D^* + {}^1A \longrightarrow {}^1D + {}^1A^*$$

   b. **Triplet–triplet:**

$$^3D^* + {}^1A \longrightarrow {}^1D + {}^3A^*$$

   These processes are analogous of (2) and (3), and they may be considered as occurring via short-lived excimer/exciplex intermediates. In principle, electron-exchange processes can be distinguished from collisional ones when the latter are inhibited in rigid or highly viscous systems.
5) Radiationless migration due to Coulombic interactions

$$^1M_A^* + {}^1M_B \longrightarrow {}^1M_A + {}^1M_B^*$$

   The corresponding triplet–triplet migration process is forbidden because of the low absorption intensity of S_0–T_1.
6) Radiationless transfer due to Coulombic interactions
   a. **Singlet–singlet:**

$$^1M^* + {}^1Y \longrightarrow {}^1M + {}^1Y^*$$

   b. **Triplet–triplet:**

$$^3M^* + {}^1Y \longrightarrow {}^1M + {}^3Y^*$$

Again singlet–triplet and triplet–triplet transfers are negligible because of the low absorption intensity of S_0–T_1.

The conditions for Coulomb-mediated energy transport, (5) and (6), are similar to those for radiative migration and transfer (1), namely, an overlap of the donor emission with the acceptor absorption spectrum and an allowed transition in the acceptor. The condition for electron-exchange interactions, (4), are similar to those for collisional transport (2), (3), namely, short-range interaction between the donor and the acceptor and spin conservation in the transition. Singlet–singlet migration and transfer can occur radiatively, collisionally, by electron exchange, and by Coulomb interaction. Triplet–triplet migration and transfer can occur collisionally or by electron-exchange interaction. Triplet–singlet transfer can occur radiatively or by Coulomb interaction.

When the acceptor molecule is in the triplet ground state (e.g., $^3O_2$) or in an excited triplet state $^3A^*$, additional migration and transfer processes are possible:

$$^3D^* + {}^3A \longrightarrow {}^1D + {}^1A^*$$
$$^3D^* + {}^3A^* \longrightarrow {}^1D + {}^3A^{**}$$

These reactions may occur because of encounter complex (excimer/exciplex) formation and dissociation. The first reaction leads, for example, to the formation of singlet oxygen, highly reactive species and a killer of biomolecules and living cells. It can take place forming an intermediate excimer/exciplex in the singlet state $^1(M \cdot Y)^*$. Remember that two $S = 1$ spin vectors can compose into a state with overall spin value $S = 2, 1, 0$, with multiplicity 5 (quintet), 3 (triplet), 1 (singlet), respectively.[1)] So, of nine states only one has singlet character. In other words, the probability to form a singlet encounter complex from two triplets is 1/9.

6.3.1 Radiative Migration

Radiative energy migration manifests itself as reabsorption of the emission, giving rise to a reduced emission quantum yield. If "a" is the probability of self-absorption, $(1 - a)$ is the photon escape probability. Photons that are absorbed, however, can be reemitted, and this process can repeat virtually infinite times. The measured quantum yield H due to the escape of fluorescence is

$$H = \eta F(1 - a)\left[1 + a\eta_F + a^2\eta_F^2 + \cdots\right] = \frac{\eta_F(1 - a)}{1 - a\eta_F} \tag{6.5}$$

where $\eta_F = \int \phi(\nu)d\nu$ is the fluorescence quantum yield of the isolated molecule. The series in square brackets corresponds to the emission after successive escape and recapture of photons for 1, 2, 3, ... times. Accordingly, the lifetime of the excited state is also affected, because the radiative rate is reduced (or the radiative time enhanced) by $\tilde{K}_R = (1 - a)k_R$, and thus the total decay rate of the state is

$$\tilde{K} = \tilde{K}_R + k_{NR} = (1 - a)k_R + k_{NR} = (1 - a\eta_F)\, k \tag{6.6}$$

The experimental consequences of self-absorption (or radiative energy migration) are

1) The reduced quantum yield of emission $H = \frac{\eta_F(1-a)}{1-a\eta_F}$.
2) The reduced excited-state lifetime $\tilde{K} = (1 - a\eta_F)\, k$.
3) The change in PL spectral shape, where the high-energy region of the spectrum (typically the 0–0 vibronic transition) is depressed.

1) The combination of two spin states, S_1 and S_2, follows the rules for angular momenta in quantum mechanics. The possible outcomes are $S = |S_2 + S_1|, |S_2 + S_1 - 1|, |S_2 + S_1 - 2|, \ldots, |S_2 - vS_1|$; with multiplicity $2S + 1$.

The probability of self-absorption depends on the overlap between emission and absorption spectra, the sample thickness, and molecular concentration. Clearly, a neat homogeneous ensemble of identical molecules will have a very small reabsorption probability, because such an overlap will be small. In the presence of disorder, however, this overlap will tend to increase. In terms of all those parameters the probability of reabsorption can be written as

$$a = \frac{1}{\eta_F} \int \phi(\nu)(1 - 10^{-\varepsilon(\nu)[M]L})d\nu \approx \frac{2.303\,[M]\,L}{\eta_F} \int \phi(\nu)\varepsilon(\nu)d\nu \qquad (6.7)$$

where integration is extended on the whole frequency range, [M] is the molecular concentration in moles per cubic decimeter and L is the sample thickness. Obviously the strategy for reducing self-absorption is to use a thin sample at very diluted concentration.

When a guest molecule in solution is responsible for energy transfer, the same expression for a is approximately valid provided the concentration and the extinction coefficient is that of the guest.

6.3.2
Nonradiative Energy Transfer

Let us consider a donor molecule D and an acceptor molecule A at distance r_{AD} in a rigid solvent whose lowest excited state lies well above the molecular optical gaps. D and A can be the same or different molecules. Energy deposited in molecule D can be transferred nonradiatively to molecule A by Coulomb or exchange coupling. The intermolecular transfer is a resonance process that involves isoenergetic radiationless transitions in D^* and A. The possible $D^* \rightarrow D$ transitions correspond to the donor fluorescence spectrum $\phi_D(\lambda)$ and occurs in various vibrational levels of D. The possible $A \rightarrow A^*$ transitions correspond to the acceptor absorption spectrum, $\varepsilon(\lambda)$, and occurs in various vibrational levels of A^*. The energy transfer is adiabatic, the difference $E_D - E_A$ in electronic energy between D and A is partitioned as vibrational energy between the two final species D and A^*. Normally the transfer is slow with respect to vibrational relaxation, corresponding to very weak intermolecular coupling, so that the process is irreversible and can be described by time-dependent perturbation theory. This is similar to Kasha rule for emission: the system will first relax to the lowest excited state in D before transferring energy to A. Owing to very weak coupling, exciton delocalization does not take place and ground-state absorption is unaffected. The energy transfer rate k_{DA} depends on the spectral overlap, J, between donor absorption, $\varepsilon(\lambda)$, and acceptor fluorescence $\phi_D(\lambda)$.

$$J = \int \phi(\lambda)\varepsilon(\lambda)\lambda^4 d\lambda \qquad (6.8)$$

where the extinction coefficient is usually expressed in liters per mole per centimeter. Note that the fluorescence spectrum ϕ is that in the absence of the acceptor and is defined as $\eta_F = \int \phi(\nu)d\nu$, where η_F is the fluorescence quantum yield. The spectral overlap J determines the density of equienergetic $D^* \rightarrow D$ and $A \rightarrow A^*$

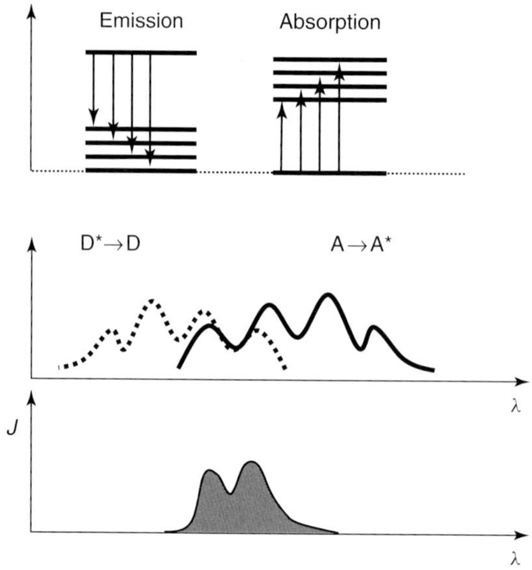

Figure 6.6 The spectral overlap J in the theory of energy transfer.

transitions, as schematically shown in Figure 6.6. In the very weak coupling limit the rate of energy transfer is worked out through the Fermi golden rule for perturbation theory:

$$k_{\mathrm{DA}} = \frac{2\pi}{\hbar}\beta^2\rho_{\mathrm{E}} \tag{6.9}$$

where β is the electronic interaction energy and ρ_{E} is the density of states, related to spectral overlap J. The interaction energy β can be due to Coulomb coupling or exchange interaction. To see why there are two possible mechanisms we start writing the initial state wavefunction as a product

$$\Psi_{\mathrm{D^*A}} = \frac{1}{\sqrt{2}}\left[\phi_{\mathrm{D}}^*(1)\phi_{\mathrm{A}}(2) - \phi_{\mathrm{D}}^*(2)\phi_{\mathrm{A}}(1)\right] \tag{6.10}$$

We introduce the information on the electrons, assuming that only two electrons – one on D and one on A – are involved in the transition. $\Psi_{\mathrm{D^*A}}$ is the properly antisymmetrized total wavefunction with respect to electron exchange of the system (D . . . A) before the transition. After the transition the final wavefunction is

$$\Psi_{\mathrm{DA^*}} = \frac{1}{\sqrt{2}}\left[\phi_{\mathrm{D}}(1)\phi_{\mathrm{A}}^*(2) - \phi_{\mathrm{D}}(2)\phi_{\mathrm{A}}^*(1)\right] \tag{6.11}$$

Note also that here we do not build a coherent superposition of excited-state wavefunctions, for in the antysimmetrized product the excitation is first in D and later in A. This means we are considering a noncoherent process of energy transfer between D and A. The rate of transfer, being described by the Fermi golden rule for perturbative theory contains the resonance energy term

$$\beta = \langle\Psi_{\mathrm{D^*A}}|\, V\, |\Psi_{\mathrm{DA^*}}\rangle \tag{6.12}$$

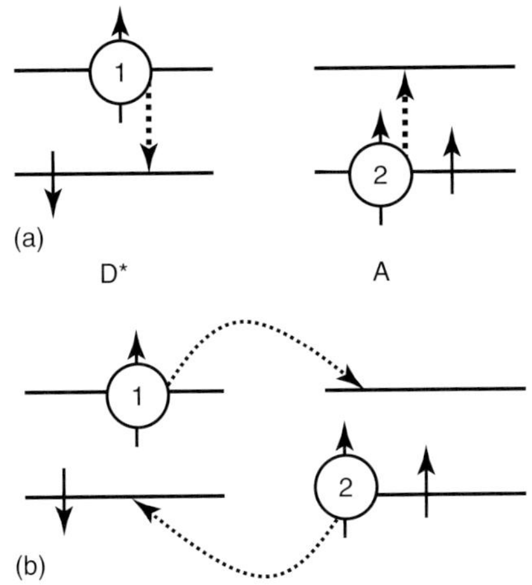

Figure 6.7 (a) Forster dipole coupling model for energy transfer. (b) Dexter electron-exchange model for energy transfer.

where V is the Coulomb interaction that causes the transition. β has two terms, the Coulomb term

$$\beta_C = \langle \phi_D^*(1)\phi_A(2) | V | \phi_D(1)\phi_A^*(2) \rangle \tag{6.13}$$

and the exchange term

$$\beta_E = \langle \phi_D^*(1)\phi_A(2) | V | \phi_D(2)\phi_A^*(1) \rangle \tag{6.14}$$

These two mechanisms are depicted in Figure 6.7.

We first discuss the Coulomb coupling mechanism (also (5) and (6) above), proposed by Förster and Oppenheimer. The Coulomb interaction can be expressed by the multipole expansion. Limiting our discussion to the leading dipole–dipole term, the interaction energy is expressed as a function of the transition dipole moments of the donor and acceptor molecules according to

$$\beta \propto \frac{\overline{\mu}_D \cdot \overline{\mu}_A}{r_{AD}^3} \tag{6.15}$$

The rate K_{DA} depends on the square of the interaction term, thus showing the peculiar r^{-6} distance dependence. Combining previous equations Förster obtained

$$k_{DA} = \frac{1}{\tau_D}\left(\frac{R_0}{R}\right)^6 \tag{6.16}$$

where τ_D is the excited-state lifetime of the donor, $\tau_D^{-1} = k_D = k_R + k_{NR}$, in the absence of the acceptor. R_0 is the critical transfer distance at which the energy transfer rate is equal to the deactivation rate of the isolated molecule. In other words, at this distance, 50% of the population undergoes energy transfer and the fluorescence yield from the donor is halved. The critical transfer distance is given by

$$R_0 = 0.211\left[k^2 n^{-4} J(\lambda)\right]^{1/6} \tag{6.17}$$

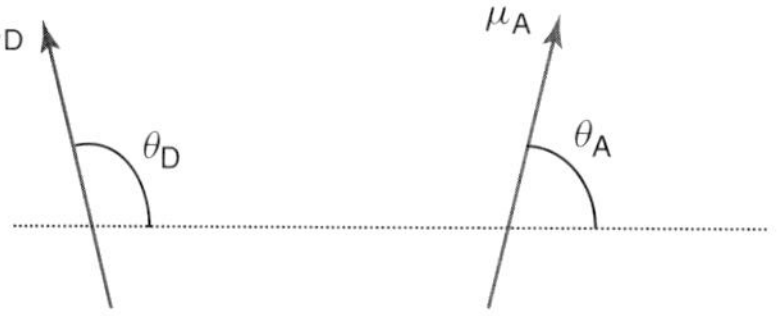

Figure 6.8 The dipole coordinates for describing the dipole coupling.

where "n" is the refractive index, J is the spectral overlap, and k is a geometrical factor, $\kappa = \cos\theta_{\mu_D\mu_A} - 3\cos\theta_D\cos\theta_A$ (Figure 6.8) accounting for dipole–dipole orientation. For isotropically distributed dipoles $k = 2/3$. Sometimes a normalized, unit-area fluorescence spectrum, $F(\lambda)$, is used in the spectral overlap definition, and the PL quantum yield compared specifically in the definition of R_0 according to $\phi(\lambda) = \eta_F F(\lambda)$.

The efficiency of energy transfer is worked out simply from the probability ratio as

$$\eta_{ET} = \frac{k_{ET}}{k_{ET} + \frac{1}{\tau_D}} = \frac{1}{\left(\frac{R}{R_0}\right)^{1/6} + 1} \tag{6.18}$$

A rate of energy transfer of the kind $k = \frac{C_1}{R^6}$ leads to a peculiar time dependence of the donor population. Note that the acceptor can be the same molecule, for instance, in the process $S_1 + S_1 \rightarrow S_n + S_0 \rightarrow S_1 + S_0 + \Delta$, which depicts a bimolecular singlet annihilation reducing the number of singlet states (Δ be a phonon energy term), or $S_1 + S_1 \rightarrow S_n + S_0 \rightarrow P^+ + P^- + S_0$, which regards the further singlet fission into polarons. In both processes the singlet population rate equation contains a term such as $\frac{dN}{dt} = -\gamma N^2$. The event probability "$p(t)$" will reduce while annihilation takes place, because the number of acceptor molecules is reduced in time. A simple way to account for this is to write $\frac{dp}{dt} = -\frac{C_1}{R^6}p(t, R)$. The corresponding probability is thus exponential and should be weighted by the molecular density at distance R:

$$N \times p(R, t) = N \times C_2 R^2 \exp\left(-\frac{C_1}{R^6}t\right) \tag{6.19}$$

The total rate coefficient, that is, the probability per unit time, is dp/dt integrated over the whole range of distances,

$$\begin{aligned} \frac{dN}{dt} &= -\left(\int_0^\infty k\,dR\right) N = -N^2 \int_0^\infty \frac{d}{dt}\left[C_2 R^2 \exp\left(-\frac{C_1}{R^6}t\right)\right] dR \\ &= N^2 \int_0^\infty C_2 R^{-4} \exp\left(-\frac{C_1}{R^6}t\right) dR = C_3 t^{-\frac{1}{2}} N^2 = \gamma(t) N^2 \end{aligned} \tag{6.20}$$

This shows that the process is dispersive, or non-Markovian, that is, with a time-dependent probability. This is true even for the homogeneous initial distribution because nearby pairs will annihilate first, while distant pairs will become longer lived. The constant C_3 can be represented in terms of physical quantities to be related to the model, $C_3 = \left(2\pi^{3/2}/3\right) R_0^3 \tau_F^{-\frac{1}{2}}$, where R_0 is the critical radius,

τ_F the radiative decay time and an average geometrical factor is included. If the states can move, the diffusion-controlled annihilation process in which two of them react only if their distance is below R_{AD} is again time dependent, with $\gamma = 8\pi DR_{AD}\left[1 + R_{AD}\left(2\pi Dt^{-1/2}\right)\right]$, where D is the diffusion constant. Using typical values for conjugated polymers the time dependence can be dropped after 1 ps.

Finally, the mechanism of mutual annihilation between fixed species and annihilation of mobile species can both take place, leading to a combined bimolecular annihilation probability $\gamma = 4\pi 0.676 D^{3/4} R_0^{3/2} \tau_F^{-1/4} + C_3 t^{-1/2}$.

Dexter has extended the Förster analysis to a higher multipole–multipole and electron-exchange interactions. When donor–acceptor interaction occurs at close distance, before the full quantum mechanical regime sets in, retaining quadrupole and higher terms in the expansion might help. Multipole interactions decrease as higher inverse power of the intermolecular distance. These interactions are thus of a range shorter than dipole–dipole interactions. Yet whenever a symmetry forbidden transition is involved they may have a role. In benzene, for instance, octupole–octupole interaction is dominant. In plants and bacteria in which carotenoids have light harvesting function, energy is transferred also from the lower lying, dipole forbidden state of the carotenoid to chlorophyll molecules. In this case, the role of higher terms in the multipole expansion has been invoked.

When the donor and acceptor transitions are spin forbidden, the electron-exchange (Dexter) interaction is dominant. Here wavefunction overlap is essential to allow electron transfer. The probability is still dependent on the spectral overlap J but independent of the optical transition moments.

$$k_{DA}\,(\text{exchange}) = Be^{-2\frac{R_{DA}}{L}} J \tag{6.21}$$

The exponential decay on intermolecular distance stems from the required wavefunction overlap. B is a coupling constant that cannot be directly related to optical properties.

6.4
The Vavilov–Kasha Rule

Both Vavilov and Kasha statements regard a generally observed behavior in photoexcited molecules, but they are not fundamental laws. In short, the two statements are

1) Kasha:
   Fluorescence is observed exclusively from the lowest electronic excited state.
2) Vavilov:
   Fluorescence quantum efficiency is independent of the excitation wavelength (for non ionizing radiation).

Both statements indirectly rely on the fact that nonradiative decay is extremely fast, and indeed much faster than radiative (fluorescence) recombination. This is well accounted for by the gap law (see below), essentially linking the decay rate to the interstates gap. Higher lying excited states will find very small gaps in energy with lower states. The larger gap is found between S_1 and S_0, thus from S_1 luminescence may take place. For isolated molecules these rules are fairly well fulfilled. The Vavilov rule is stricter, but for this reason less general. In large organic molecules, long-chain polymers, or aggregates and solids it might fail because high-energy excitations may open up decay paths not leading to the lower singlet state (for instance, charge generation and singlet fission). As a consequence, the population of S_1 and the emission yield will be reduced on increasing the excitation energy (above each threshold for alternative paths).

The Kasha rule holds true in many more situations, and notably also in disordered solids in which emission is seen from a subset of low-lying energy states. In disordered solids a localization threshold marks the border below which energy migration does not take place, and "resonant" emission occurs (Figure 6.9). Above this threshold there are energy states that are efficiently coupled to lower energy states, and emission occurs after relaxation. This explains why the emission lineshape is often narrower and more structured then the absorption one. Excitation below this energy reaches isolated states that decay without energy transfer. Once again it is the balance between the rates that determine the overall outcome. If the energy transfer rate is faster than the radiative decay rate, most of the energy will be reemitted after relaxation; if not, emission will occur from the absorbing state. As a typical order of magnitude, energy transfer times above localization are in 100 fs

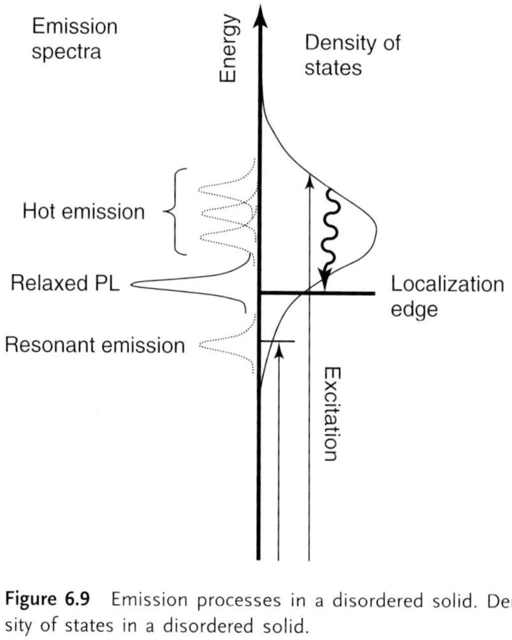

Figure 6.9 Emission processes in a disordered solid. Density of states in a disordered solid.

to 1 ps time scale, while radiative decay for allowed transitions is in the 1–10 ns time scale. For this reason, hot emission is extremely weak or nonexistent.

Experimental probes of such processes are time-resolved PL, pump probe measuring stimulated emission, excited-state absorption, or polarization memory (anisotropy) decay. Excitation profiles (i.e., emission vs excitation energy) do provide information about this as well, but in a less direct way. Intramolecular relaxation may be confused with intermolecular relaxation.

6.5 The Gap Law and Radiationless Transitions in Molecules

A molecule initially prepared in an excited state will return back to its ground state after a certain time. This may occur by radiative decay, which implies the emission of a photon. Very rarely, the energy of the emitted photon is equal to the energy of the initial molecular excited state. Most often the emitted photon energy is smaller than that of the initially prepared molecular state. In addition, considering an ensemble of molecules, most of the time only a fraction of the initial energy is reemitted into visible or near-infrared photons. Emission in the far-infrared is usually not measured. The question that arises first is where the energy went. For isolated systems, energy conservation predicts that only intramolecular redistribution is possible. Theory also states that within a suitable time the molecule will return to the initially prepared state (recurrence time), although we have not noted this in our experience. The reason is that molecules are never isolated. Energy initially deposited into a well-defined molecular excited state ends up in the environment, and for statistical reason (the huge number of states in the bath) this energy will never come back. Essentially what we say is that, after some time any molecule will dissipate excess energy, respect to thermal equilibrium, by transferring to the environment.

The deactivation rate of an electronic excited state is written as a sum of rates $k = k_{\mathrm{R}} + k_{\mathrm{NR}}$, where R denotes radiative, that is, the emission of one photon, and NR denotes nonradiative. This leads to exponential decay of the initial state population. The fraction of energy going into decay channel "j" is given by the ratio $\eta_j = k_j/k$. From the experimental point of view, nonradiative decay can be evaluated by measuring the PL quantum yield, η_{F}, and the PL lifetime, k^{-1}. Because $\eta_{\mathrm{F}} = k_{\mathrm{R}}/k$, from these measurements one get to know both k_{R} and k, so k_{NR} can be worked out. The measurement of the PL quantum yield requires a careful calibration versus a standard and the collection of emitted light in the whole solid angle using an integrating sphere. The use of the S–B relationship is an alternative to the measurement of η_{F}, as discussed in Section 6.2. The measurement of PL lifetime can be obtained by time-resolved experiments using either a streak camera or a photon counting system. Very short PL lifetime can be measured using the up-conversion technique. The pump probe experiment can provide a more detailed picture of the deactivation process, detecting some of the nonradiative decay channel in direct way, for example, the triplet population from the initial singlet. However, probing electronic transitions (in the visible or near-IR spectral region) fail to detect vibrational energy. Using IR probing pulses one can have a less partial view of the process, yet it is virtually impossible to detect all the modes involved in the process.

Nonradiative transitions are those not involving the absorption or the emission of photons. They may occur between different excited vibronic states of the same spin multiplicity (IC), between states of different spin multiplicities (ISC), between vibrational states of the same molecule (Internal vibrational redistribution (IVR)), or between molecular and bath states (VC). They involve equienergetic initial and final states; in other terms, they should be draw horizontally in the Jablonsky diagram, as in Figure 6.10 (dashed arrow). This figure may look strange. The reason is that we plot all vibrational levels, not just those optically coupled, and stress energy conservation. Till VC does not set in, all the energy is stored in the molecule. Figure 6.11 depicts IC from S_2, after optical excitation, to S_1. Here we specify optically active and dark vibrational modes. Note that optical probing, for instance, stimulated emission, would detect initially the hot S_2 state, then the relaxed S_2 state, followed by the relaxed S_1 state, as if the system would transit from the bottom of S_2 to the bottom of S_1, which is not the real process. Radiationless transitions play a crucial role in deactivation of molecular states. They generate phonons, which represent, at equilibrium, the effective heating of the sample. A cascade of phenomena takes place from the initial phonon generation (out of equilibrium state, temperature is not defined) to thermalization (thermal phonon distribution corresponding to a temperature higher than that of the environment,

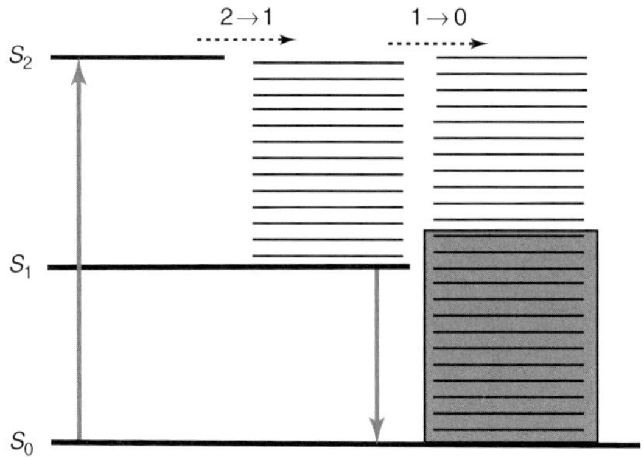

Figure 6.10 Jablonsky diagram with initial optical excitation (gray arrow). Dashed arrow shows IC. Downward gray arrow is fluorescence. The fine structure represents vibrational energy levels. In large molecules this is essentially a continuum. Nonradiative transitions are horizontal because of energy conservation. After IC, the ground state has a large vibrational energy content, equal to the S_2-S_0 energy gap. If fluorescence occurs, part of the energy is reemitted and the vibrational energy in S_0 is $E(S_2)-E(S_1)$, according to the Kasha rule (gray area). If IC only deactivates S_2, then $E(S_2)-E(S_0)$ is the vibrational energy in S_0.

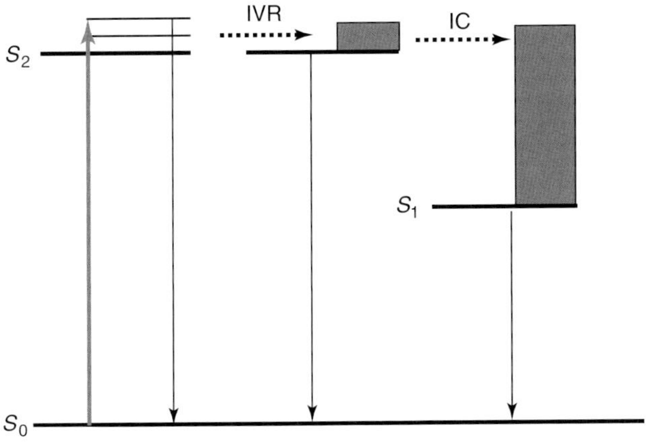

Figure 6.11 Optical excitation to S_2 is followed by IVR and then IC to S_1. Thin levels are optically active vibration. Gray area represents the nonoptical vibrational continuum. Thin downward arrows are optical probes (stimulated emission).

known as *hot phonon distribution*) to cooling with equilibrium phonon distribution at the actual final temperature of the molecular ensemble.

The picture for radiationless transitions has changed with time. In reviews of the 1970s, authors regard crossing of potential energy curves in the configurational space as "not very likely" and not necessary to explain raditionless transitions, which were described by tunneling between adiabatic states. In a recent book by Domcke *et al.* [1] the initial statement is just the opposite: intersection between potential energy surfaces are ubiquitous and play the crucial role for explaining nonradiative transitions via nonadiabatic state crossing. These crossing points are called *conical intersection* (when fulfilling certain conditions). So today, if you ask why large polyconjugated molecules have extremely fast relaxation times to the lowest excited state, the answer is because of conical intersections. Some time ago, the answer would have been because of high density of states, boosting the transition rate between adiabatic states. It is instructive to have an idea of both mechanisms in order to grasp the concept of radiationless transition. They are both active in large molecules. See Figure 6.12 for a pictorial description.

The tunneling between adiabatic states can be described by the Fermi golden rule of time-dependent perturbation theory:

$$k_{\mathrm{NR}} = \frac{2\pi}{\hbar} V_{\mathrm{fi}}^2 \rho_{\mathrm{fi}} \tag{6.22}$$

where V_{fi} is the matrix element of the nonradiative coupling term and ρ_{fi} is the density of final states.

This approach is similar to that followed for radiative transitions. In this case, however, the electric dipole operator is substituted by the nuclear kinetic energy operator or other nonadiabatic operators. These are the energy terms that we disregard in setting adiabatic approximation. The golden rule is based

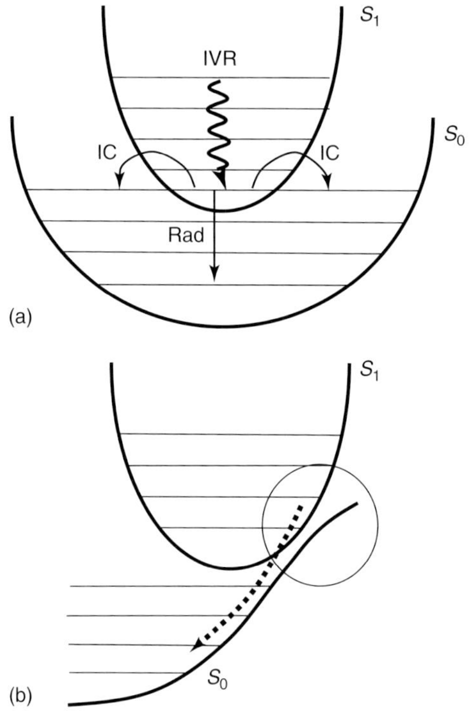

Figure 6.12 (a) IC tunneling between two adiabatic states (bent arrows). Wavy arrow shows internal vibrational redistribution. Solid arrow is decay of S_1 by photon emission. (b) Intersection of two adiabatic states leads to avoid crossing (in the circle). Dashed arrow shows nonadiabatic path for $S_1 - S_0$ IC.

on the assumption that transition is from a discrete level to a continuum of levels and the states involved are most of the time zero-order solution to the Schrödinger equation, that is, obtained by neglecting all perturbative terms. This is a further approximation because it neglects the occurrence of state mixing due to nonadiabatic perturbation. Sometimes, as in ISC, this mixing is crucial and will be considered. Another situation where nonadiabatic coupling cannot be disregarded is at potential energy surfaces crossing, where the electronic gap reaches zero. Here the adiabatic approximation fully breaks down and new states must be considered.

The term leading to IC is the nuclear kinetic energy operator applied to the electronic wavefunction ($J_N \propto \frac{\partial^2 \phi}{\partial Q^2}$) or the mix term $\propto \frac{\partial \phi}{\partial Q_k} \frac{\partial \chi}{\partial Q_k}$, that of ISC is a combination of spin-orbit coupling H_{SO} and J_N and that of IVR is the vibrational anharmonicity due to higher terms in the phonon energy expansion ($\sim \frac{\partial^3}{\partial Q_k^2 \partial Q_j}, \ldots$). All these terms lead to nonzero off-diagonal matrix elements between the initial zero-order states. As stated above, perturbation theory may include higher order description of the states, which are described by linear combination of zero-order states (state mixing). An example is spin-orbit coupling. The zero-order solution for pure spin states contains singlet, ${}^1\Phi_n$, and triplet, ${}^3\Phi_m$ states. S^2 and S_z are good quantum numbers and no cross-talk exists between the two spin manifold. Because the spin and orbital magnetic dipoles can interact in the molecule, the spin-orbit coupling term H_{SO} is not zero, and the zero-order states are not its eigenstates. The nondiagonal matrix elements $\langle {}^1\Phi_n | H_{SO} | {}^3\Phi_m \rangle$ are nonzero. A first-order correction leads to contaminated states, where singlets acquire some triplet character and triplets acquire some singlet character (see also below). This contamination due to state mixing will occur considering any perturbative term of the molecular Hamiltonian. There is then a certain degree of arbitrariness in choosing the basis for describing the phenomena. For IC, contaminated states are not important, as it is for IVR. First-order time-dependent perturbation theory only including J_N allows accounting for it. For ISC, however, only spin contamination allows a simple explanation of phosphorescence (the radiative decay of the triplet $T_1 - S_0$, strictly forbidden for pure states), and the description of radiationless ISC should contain both H_{SO} and J_N, appearing in second-order expressions.

The transition matrix term for IC can be approximated by

$$H_{fi} = \langle \phi_f \chi_f(\Delta E) | \hat{J}_N | \phi_i \chi(0) \rangle \tag{6.23}$$

where adiabatic approximation is used by adopting the factorization of the total wavefuction in the electronic and vibrational parts. The basic process is depicted in Figure 6.13. The initial state is the vibrationally relaxed vibronic state $\phi_i \chi(0)$. Owing to IC, the initial state is transformed into the lower lying vibronic state $\phi_f \chi_f(\Delta E)$, which is a combination of electronic (E_f) and vibrational energies (ΔE) that allows fulfilling energy conservation: part of the initial electronic energy is thus transformed into vibrational energy. In the golden rule equation, tunneling through a potential barrier is replaced by a Franck–Condon (FC) overlap factor, and the nonradiative rate $k_{nr} = \frac{2\pi}{h} J_{fi}^2 FC \rho_{\Delta E}$, is obtained from

$$k_{nr} \propto |H_{fi}|^2 \rho_{\Delta E} \cong \left| \langle \phi_f | \hat{J}_N | \phi_i \rangle \langle \chi_f(\Delta E) | \chi(0) \rangle \right|^2 \rho_{\Delta E} \tag{6.24}$$

The FC factor is a product of FC overlap integrals:

$$FC = \sum_p P \left[\prod_{k=1}^{N} |\langle \phi_{f,n}(n_k) \mid \phi_{i,k}(0) \rangle| \right]^2 \tag{6.25}$$

and following energy balance the disposed energy ΔE goes into phonons according to a proper distribution of population factors n_k such as $\sum_k n_k \hbar \omega_{f,k} = \Delta E \pm \frac{1}{2\rho_{\Delta E}}$.

Figure 6.13 Internal conversion from a vibrationally relaxed initial state "I" to a final state "f" with electronic and vibrational energy.

Here ρ_{Δ_E} is the state density, defined as the reciprocal width of a single vibrational level. The operator $\sum_{p} P$ permutes the vibrational quantum numbers n_k among the N normal modes subject to the energy conservation rule. As discussed before, both displaced (different equilibrium position) and distorted (different vibrational frequency) harmonic oscillations lead to non-null FC integrals. Because in those expressions there is a huge number of vibrational modes involved, the density of states is sufficiently large that the FC factor has no vibrational structure and it is a monotonically decreasing function of energy ΔE. The density of states $\rho_{\Delta \mathrm{E}}$ (i.e., the number of suitable combinations leading to energy conservation) is increasing with energy ΔE. The product FC $\times\ \rho_{\Delta E}$ is, however, decreasing with energy, in an exponential manner. This poses the basis for the gap law, as discussed below. The symmetry selection rules of the transition matrix become unimportant because of the large number of combinations that adds up, providing a sizable contribution also for very small electronic terms. As a consequence, orbital symmetry characters, such as even (g) and odd (u) do not play a crucial role, contrary to what happens for dipole allowed transitions. In addition, vibronic activation allows nonzero coupling even between strictly forbidden states. It is interesting to distinguish between the two roles of the vibrational modes: accepting modes are those involved in the FC term that receive and store the excess energy given away by the initial electronic state; promoting modes are those involved in the vibronic term of the matrix element. Suppose the matrix element $\langle \phi_\mathrm{f} | \hat{J}_\mathrm{N} | \phi_\mathrm{i} \rangle$ is zero because of symmetry, the mix term $\langle \phi_\mathrm{f} | \frac{\partial}{\partial Q_k} | \phi_\mathrm{i} \rangle \left\langle \chi_k^\mathrm{f} \middle| \frac{\partial}{\partial Q_k} \middle| \chi_k^\mathrm{i} \right\rangle \neq 0$ if the vibrational mode "k" has proper symmetry.

The quantum mechanical expressions above have the mere scope to suggest the physical origin of nonradiative decay and highlight the possible ingredients. To sum up, it is the nuclear motion that brings the molecule into an equienergetic state with a highly vibrationally excited, lower electronic energy state, thus allowing a transition.

In summary, look again at Figure 6.11, which reports the most relevant phenomena in spectroscopy. Following excitation, a vibronic state is populated, where energy is both electronic and vibrational: $E_\mathrm{i} = E_{S_2(0)} + n_k \hbar\omega_{2,k}$ (here zero point energy is in the electronic term, and only one vibrational mode, "k," is coupled to the transition). IVR leads to a redistribution of the initial vibrational energy to many other modes, not necessarily coupled to electronic transitions. Those states are not detected and may be considered "dark." A spectroscopic probe would detect the relaxed electronic state, $S_2(0)$, even if excess energy is still in the molecule and not yet fully thermalized. Then IC leads to population of the lower lying state, $S_1(0)$, and a combination of vibrational levels, according to: $E_{S_2(0)} = E_{S_1(0)} + \sum_k n_k \hbar\omega_{1,k}$.

Here we assume that none of the involved final vibrational states are optically active. This is indeed a common situation, simply because of statistical reasons: there can be hundreds of vibrational modes in large molecules ($3N - 6$ where N is the number of atoms) and only few are optically active. The kinetic energy operator has a defined symmetry; however, selection rules become shaky because of the huge density of states. As a result, the final state has dark vibrational energy and

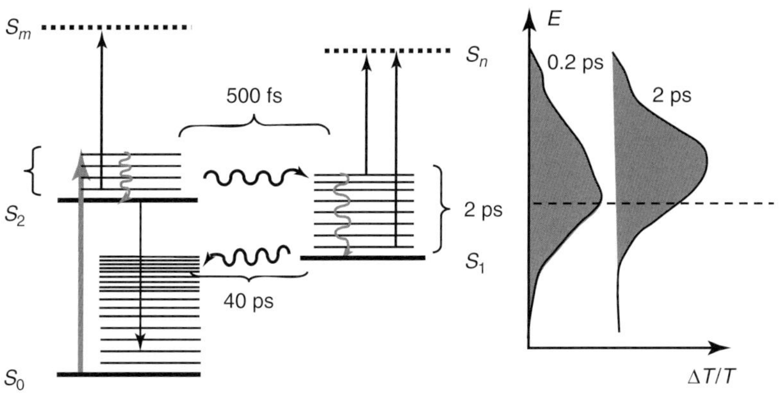

Figure 6.14 The simplified relaxation scheme in β-carotene. Gray arrow: initial optical excitation. Thin black arrows: optical probing. Wavy arrow: IC and IVR. Time scales are reported (not time constants). On the left, transient S_1-S_n absorption bands are shown for 0.2 and 2 ps pump probe delay.

appears "relaxed" to optical probes. The excess energy does not, however, disappear from the excited molecule, which should still be considered as hot. In some cases, the final state has a fraction of energy stored in optically active vibrational modes (total symmetric for vibronic transitions). An example is β-carotene, where S_1 is formed from internal conversion of higher lying singlet states with a fraction of energy stored in an optically active vibrational mode (Figure 6.14). In this case, excited-state absorption allows monitoring part of the IVR process, associated with blueshift of the S_1-S_n absorption band and reshaping, as the population relaxes toward the bottom of the vibrational manifold.

The complex expression for nonradiative decay rate, k_{nr}, as reported above, can be approximated by the so-called gap law, as proposed by R. Englman, and J. Jortner in 1970 [2]. In a simplified version, valid in the strong coupling regime (i.e., when the potential energy surfaces are well displaced between the initial and final states along few vibrational modes), the nonradiative decay rate can be written as

$$k_{\text{nr}} \approx A \exp\left(-\frac{|\Delta E|}{B}\right) \tag{6.26}$$

In the formula, A is a pre-factor, not a priori known and dependent on the coupling strength, B is an average phonon frequency or relaxation energy, and ΔE is the gap between the electronic origins. As we stated above, this approximation stems from the observation that on increasing the energy gap the FC over integrals tend to diminish faster than the increasing in density of state. This holds true only when the energy gap is large compared to the relevant vibrational energy quanta.

There are several experimental proofs for the validity of the gap law. For instance, the dependence of the phosphorescence decay rate constants on the triplet state energy in aromatic hydrocarbons, as found back in 1966. In that case, it is known that the decay is predominantly nonradiative. Another example where the gap law works satisfactorily is the S_1-S_0 decay in polyenes, for which the best parameters are $A = 3.90 \times 10^{15}\ s^{-1}$ and $B = 1353\ cm^{-1}$. The gap law provides a ground for explanation of the Vavilov–Kasha rule because any higher lying state S_n will find a ladder of lower states to decay to, with energy gap smaller than S_1-S_0. As stated above, conical intersections are considered to play a crucial role in the relaxation of large molecules. Conical intersection (CI) arises by crossing of PES. Avoided crossing due to coupling hamiltonian maybe small, and approaching the minimum gap between the energy surfaces, a complete break down of the adiabatic approximation may occur. In this situation, the scenario we reported above is no more valid. The description of the relaxation process cannot be done using perturbation theory, and the full quantum mechanic treatment becomes extremely complex. Relaxation of the electronic energy is accompanied by nuclear motion and is not separated. Indeed electronic energy is transferred to the nuclear motion during relaxation, as if the vibrational wavepacket were following a classical trajectory. Surface hopping algorithms originally developed by Tully and coworkers [4] describe the nonadiabatic coupling in between classical propagation, allowing the building of a diabatic path as reproduced by the dashed line in Fig. 6.12b. A simplified mathematical description of this phenomenon, for constant coupling between diabatic potential energy curves, is the Landau-Zener formula. Original Landau-Zener was derived for tunneling for the avoided crossing case; recently, it was extended to the CI case (see Ref. [5]). The detection of exceptionally fast radiationless decay processes appears at present to be the only way to establish by purely experimental means the existence of a CI. This, however, misses to provide a number of importantly details about the process, and most importantly conceals the undergoing nuclear dynamics. In general, advanced theoretical calculations (e.g., multiconfiguration interaction approaches coupled with nonadiabatic nuclear dynamics) are needed to assess a more complete description of the process at CI. Important information is, for instance, the involved set of coordinates associated with a conformational geometry supporting the CI. CIs in large polyatomic molecules are often associated to biradical forms, which bring about the twisting or the puckering of the molecular backbone.

6.6 Rate Equations

The time evolution of the excited-state population is conveniently described using rate equations.

As an example we write

$$\frac{dN_{S_1}}{dt} = G(t) - f(N_{S_1}) \tag{6.27}$$

Depending on the generation (G) and recombination (f) terms, this equation can be solved numerically to yield $N_{S_1}(t)$.

The generation term can be

$$G(t) = \sigma_{01}(\omega_{\text{pump}})F(t)N_0(t) \tag{6.28}$$

when optical excitation leads to S_1 population, or

$$G(t) = +\eta\frac{dN_k}{dt} \tag{6.29}$$

when S_1 is populated by relaxation from other states "k." N_0 is the ground-state population and $F(t)$ is the pump photon fluence (photon per area and time).

The recombination term can be expanded as

$$f = -\left(k_m(t)N_{S_1} + \gamma(t)N_{S_1}^2 + \cdots\right) \tag{6.30}$$

where the first term is the monomolecular decay and the second term is the bimolecular one. Other terms can contain products of different populations, such as charge states (or doublet in spin multiplicity) and triplet states, which interact and annihilate singlets.

For constant monomolecular recombination rate the solution is a single exponential decay:

$$N_{S_1} = N_{S_1}(0)\exp(-k_{m0}t) \tag{6.31}$$

where $k_{m0} = k_0 + \sum_{i=1}^{p} k_i$ is the sum of radiative and nonradiative decay rates. In many real-life situations, however, the observed population decay is not a single exponential. Here a kinetic analysis may help in assignment. A sequence of well-defined and separated time constants may come from a set of corresponding decay phenomena, each with its own characteristics. Experiments should highlight each decay mechanism separately, for instance, looking at spectral region where only one decay appears or using external perturbation or measuring conditions (temperature, light intensity, etc.) to single them out. Alternatively, the nonexponential kinetics can be due to a single process, yet dispersive, that is, with a time-dependent rate. The latter can be due to a variety of causes, and it can also be that it represents an average of a very complex situation. For instance, in the presence of disorder a superposition of hundreds of slightly different time constants will lead to an overall nonexponential decay. There are several algebric expressions for a nonexponential decay, such as power law, logarithmic, multiexponential, and so on. One that is very common and suitable for general fitting is the stretched exponential:

$$N_{S_1} = N_{S_1}(0)\exp\left\{-\left(\frac{k_0}{a}t\right)^a\right\} \tag{6.32}$$

which derives from a time-dependent rate $k_m = k_0 t^{a-1}$. Here, k_0 and "a" are parameters without a specific physical meaning.

The bimolecular decay $f = -\gamma(t)N_{S_1}^2$ with time-independent constant $\gamma(t) = \gamma_0$ (the rate here is always time dependent being a function of transient population) leads to

$$N_{S_1} = \frac{1}{N_{S_1}(0)^{-1} + C_0 t} \tag{6.33}$$

While again, in the presence of time-dependent coefficient such as $\gamma = C_3 t^{-\frac{1}{2}}$ one gets

$$N_{S_1} = \frac{1}{N_{S_1}(0)^{-1} + 2C_3 t^{\frac{1}{2}}} \tag{6.34}$$

The one over square root dependence in time finds a number of explanations, as, for instance, annihilation among fixed object at variable distance. This leads to the initial fast decay, which slows down in time, on reduction of close pairs.

In more general terms, an equation such as

$$\frac{\mathrm{d}N}{\mathrm{d}t} = -\gamma_M N^M \tag{6.35}$$

leads to an intensity-dependent decay described by

$$N(t) = \frac{N(0)}{\left(1 + \frac{1}{M-1} C_M N(0)^{M-1} t\right)^{M-1}} \tag{6.36}$$

This is the general expression of which Eq. (6.34) is the particular case $M = 2$. Higher M implies a larger number of interacting particles. In particular, in semiconductor physics the Auger recombination is a three-particle process in which an e–h pair disappears, giving away its energy to an electron (hole), which will finally be thermalized by phonon emission.

The first process (time-independent coefficient) is Markovian, that is, the decay rate is the same for equal density of states. The second process, with a time-dependent coefficient, is non-Markovian, that is, the rate of decay (or probability) depends not only on the population number but also on the time from beginning. The same population reached earlier or later will provide a different decay rate. Apart from the mathematical subtleness that implies a probability rate depending on the history of the system, there is a practical consequence: this kind of kinetics cannot be properly reproduced by numerical simulations with finite pulse excitation. In this case, the initial time is not well defined, being described by the pulse temporal envelope. Only for a δ-like excitation ($F(t) = F_0\delta(t)$) simple numerical solution can be obtained.

6.7 Triplet Generation

Dipole allowed optical transitions fulfill the rule $\Delta S = 0$. Because most molecules have singlet ground state, this implies that absorption generates singlet excited states. One may then come to the conclusion that triplet states are not important in photophysics and photochemistry. This is absolutely wrong, for triplets can be formed from photoexcited singlets and by virtue of their long lifetime they can play a crucial role in all dissipation and relaxation processes. In addition, triplets are keen to react with oxygen. This is detrimental for the hosting material because of singlet oxygen formation on energy transfer from triplet states.

ISC regards transitions between different spin manifolds. In small, isolated molecules, ISC can be radiationless (phonon assisted), assigned to spin flip, $S_n \rightarrow T_m + \Delta E$ (where ΔE is vibrational energy) or radiative (photon assisted), called *phosphorescence*, which is the emission from a triplet state, $T_1 \rightarrow S_0 + h\upsilon$.

Oxygen reacts with triplet excited states, according to the energy transfer reaction $T_1 + {}^3O_2 \rightarrow S_0 + {}^1O_2$, which leads to a reduction in triplet population. For this reason, phosphorescence quenching is used as a probe for oxygen concentration, or vice versa, oxygen quenching is used to assign a triplet state.

In molecular aggregates or solid state, there are two additional mechanisms for triplet generation: (i) radical pair recombination and (ii) singlet fission (SF).

Point (i) Regards the recombination of two oppositely charged doublets, (ii) regards the spontaneous breaking of a singlet state into two triplet states. Strictly speaking neither (i) nor (ii) is ISC, because both the initial and the final states have overall the same spin multiplicity. However, we briefly discuss them here because they are important mechanisms of triplet generation with implications in OLEDs and photovoltaic cells.

6.7.1 Spin Flip

A molecular singlet excited state may spontaneously convert into a triplet excited state. The physical origin of this violation of the $\Delta S = 0$ rule is the interaction between the angular and spin magnetic moments in atoms and molecules. The spin-orbit coupling constant, which determines the strength of the interaction, is proportional to the fourth power of the atomic number, $\zeta \propto Z^4$; as a consequence, spin-orbit coupling effects are very much larger in molecules with heavy atoms.

Owing to spin-orbit interaction and according to first-order perturbation theory, triplet states acquire a small component of singlet character and singlet states acquire a small component of triplet character. The perturbed wavefunction of S_0 is

$$\left\langle \tilde{S}_0 \right| \cong \langle S_0| - \sum_k \alpha_k \langle T_k| \tag{6.37}$$

where T_k are unperturbed, pure triplet states, and the coefficient α_k is given by

$$\alpha_k = \frac{\langle T_k| H_{SO} |S_0\rangle}{{}^3E_k - {}^1E_0} \tag{6.38}$$

The magnitude of the coefficients inversely depends on the energy gap between the ground singlet state (${}^1E_0$) and all the triplet excited states (${}^3E_k$) and on the spin-orbit coupling matrix element, which in turns depends on the symmetry of the triplet states with respect to the that of the singlet and on the atomic number

The perturbed wavefunction of each of the three components of T_1 are of the form

$$\left\langle \tilde{T}_1 \right| \cong \langle T_1| - \sum_k \beta_k \langle S_k| \tag{6.39}$$

where S_k are unperturbed, pure singlet states, and the coefficient β_k is of the form

$$\beta_k = \frac{\langle S_k | H_{SO} | T_1 \rangle}{{}^1E_k - {}^3E_1} \tag{6.40}$$

with similar dependence on energy gaps and matrix element as in Eq. (6.38). In centrosymmetric molecules, the states are of even (g) or odd (u) parity. Spin-orbit coupling occurs only between states of the same parity.

Phosphorescence is a radiative transition from T_1 to S_0, which occurs between "contaminated" states with mix character. The rate of transition is proportional to the square of the dipole matrix element, μ_P, which is given by

$$\begin{aligned}\mu_P &= \left\langle \tilde{T}_1 \middle| \mu \middle| \tilde{S}_0 \right\rangle \cong \left\langle \tilde{T}_1 \middle| \mu \middle| S_0 \right\rangle = \langle T_1 | - \sum_k \beta_k \langle S_k | \mu | S_0 \rangle \\ &= - \sum_k \beta_k \langle S_k | \mu | S_0 \rangle \end{aligned} \tag{6.41}$$

Owing to the $\Delta S = 0$ selection rule for the dipole moment operator, the term $\langle T_1 | \mu | S_0 \rangle$ vanishes, the remaining terms give rise to phosphorescence. This demonstrates the role of spin character mixing in the activation of forbidden transitions.

The term ISC most of the time regards the nonradiative conversion of a singlet state to a triplet state. The rate of conversion is in the general form of nonradiative transitions:

$$k_{nr} = \frac{2\pi}{\hbar} J_{fi}^2 FC \rho_{\Delta E} \tag{6.42}$$

where the electronic term again gets amplitude from the state mixing character of the perturbed wavefunctions (J_N also follows the $\Delta S = 0$ rule):

$$\begin{aligned} J_{fi} &= \left\langle \tilde{S}_1 \middle| \hat{J}_N \middle| \tilde{T}_1 \right\rangle \cong \left\langle \langle S_1 | - \sum_k \alpha_k \langle T_k | \middle| \hat{J}_N \middle| \langle T_1 | - \sum_k \beta_k \langle S_k | \right\rangle \\ &= - \sum_k \beta_k \langle S_1 | \hat{J}_N | S_k \rangle - \sum_k \alpha_k \langle T_k | \hat{J}_N | T_1 \rangle \end{aligned} \tag{6.43}$$

By inspection of Eq. (6.43) and reminding the form of the α_k and β_k coefficients we see that both J_N and H_{SO} are responsible for ISC. The FC term accounts for the vibrational overlap factors between the initial and final states and $\rho_{\Delta E}$ is the state density, defined as the reciprocal width of a single vibrational level.

All this rationalizes the process of singlet to triplet conversion. Note that the opposite is also possible, triplet to singlet nonradiative conversion, yet in molecules it usually does not take place to S_1 because the lowest triplet excited state is lower in energy than the lowest singlet excited state due to the exchange energy term. In other words, it is a thermal activated process. IC within each manifold is much faster than ISC, so that molecules reach the lower state in the manifold before attempting a spin flip reaction. Triplet to singlet IC can, however, describe nonradiative decay of the lowest triplet excited state to the singlet ground state.

ISC is "horizontal" in the Jablonski diagram and is followed by vibrational energy redistribution and vibrational cooling (Figure 6.15). Because both optically

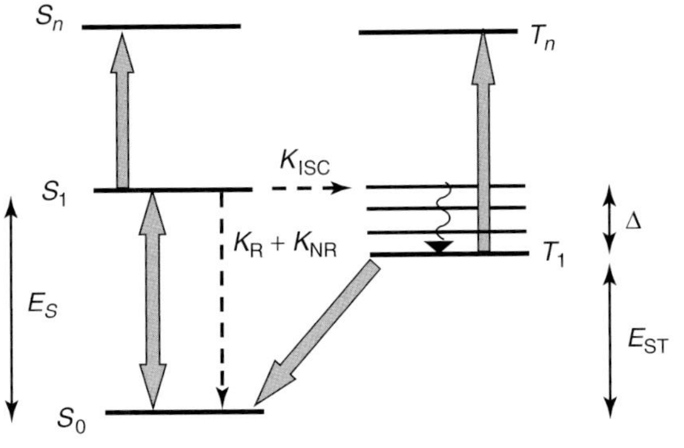

Figure 6.15 Singlet–triplet intersystem crossing. Dashed line represents transitions. Large arrows are optical transitions (double arrow represents bleaching). Thin double arrows indicate energy gaps. Δ is the singlet–triplet gap due to exchange interaction, E_{ST} is the triplet-ground-state gap, and E_S is the zero phonon ground-state absorption energy.

active and dark vibrational modes can be involved, and the second outnumber the first in large molecules, the triplet state appears immediately thermalized on optical probing.

The rate of decay of the singlet state in the presence of ISC is generally given by $k_{S_1} = k_R + k_{NR} = k_R + k_{IC} + k_{ISC}$. The ISC process quenches further the photoluminesce yield, contributing to the nonradiative decay.

The kinetics of simple ISC in a photoexcited molecule is discussed in Chapter 9.2. The deactivation rate of the triplet usually occurs on the nanosecond to microsecond time scale. The triplet population builds up exponentially at time rate k_S, that is, the triplet population forms on the same time scale of singlet decay. The yield of triplets is given by the efficiency of the process, $\eta_{ST} = k_{ISC}/k_S$. In carbon-based conjugated molecules, k_{ICS} is in the order of 1 ns^{-1}, thus what matters for getting a sizable triplet population is a comparable decay rate for the singlet, k_S. If the only deactivation channel is singlet to triplet conversion, $k_S = k_{ISC}$, then $\eta_{ST} = 1$. In general $\eta_{ST} < 1$ because the singlet decay rate is in the order of 0.01 ps^{-1} due to fast radiative and nonradiative decay channels. This explains why spin flip is not effective for higher lying singlet states. The latter have a lifetime in the order of 100 fs, and $\eta_{ST} = 10^{-4}$. Note that this tiny triplet population forms anyway in the time scale of decay of the singlet, here 100 fs. So spin flip can take place in a very short time, but with very small yield.

Suppose you do an experiment of pump probe. After excitation with a 100 fs pump pulse, you clearly see the triplet–triplet absorption band in the transient spectrum at null pump probe delay. Is this due to spin flip? You can decide by doing some simple evaluation. Let us assume that the spectrum also shows singlet–singlet absorption. This excludes spin flip from singlet, which would otherwise be depleted. In general, if singlet decay kinetics does not match triplet buildup, what you see is not spin flip. Suppose there is no singlet–singlet absorption or any other singlet feature. Possibly the singlet was deactivated, forming triplets by spin flip. You can validate this idea, making a simple estimate of triplet yield. Assume that the triplet cross section for absorption is the same as that of the singlet ground state (this might sound crude, but it works for this kind of order-of-magnitude estimate). Bleaching will then tell you how many molecules are in the triplet state. If you know how many photons were absorbed, you can estimate the triplet yield. Let us assume that it is 1%. If your system does not have heavy atoms, you can also assume that the ISC time constant is in the order of 1 ns. In 100 fs the expected yield is 10^{-4}, as we worked out before, and this again excludes spin flip as a mechanism. If, however, what you see corresponds to about 10^{-4} of the initial singlet population, then spin flip might be the right assignment.

As mentioned above, spin flip may also take place in weakly bound charge pairs (polaron-pairs). The process is schematically described by $^{1}(P^{+}\cdots P^{-}) \rightarrow {}^{3}(P^{+}\cdots P^{-})$.

The initial pair may be in the singlet state because it comes from photoexcitation either directly or on dissociation of a photoexcited singlet. Singlet and triplet states are nearly degenerate in energy and polarons in the pair can independently flip their spin, because of spin–lattice interaction or the interaction of the electron spin magnetic moment with the nuclear magnetic moment (hyperfine interaction). In π-conjugated materials this interaction is weak because π-orbitals have zero amplitude on the nuclei. As a consequence, the typical radical spin-flip rate is rather slow, in the order of 1 μs^{-1}. This has been claimed as an advantage for organic spintronic. However, the overall performance remains poor because the important figure in spintronics is the distance traveled by the spin-carrier before losing the spin information. The later is very small because of the very small mobility in organics. In addition, there are recent results challenging the notion of very weak spin relaxation, rising the radical spin flip rate to 0.1 ns^{-1}.

An external magnetic field can affect the yield of singlet and triplet in the weakly bond pair. The most amazing application of this phenomenon is the magnetic compass in bird navigation. Migratory birds possess a physiological magnetic compass, enabling them to make correct directional choices during their migratory flights. The biophysical mechanism allowing this is still unknown, but the hypothesis of the radical pair mechanism is receiving increasing attention. The mechanism is depicted in Figure 6.16. Light activates a DA pair, thereby inducing an electron transfer to create a radical pair. Singlet and triplet states coherently interconvert because of the combined effect of the hyperfine interaction and the external magnetic field. Radical pairs can recombine from the singlet state but not from the triplet. Alternatively, they can lose the coherent spin phase and decay into products. The external magnetic field, depending on both amplitude and orientation, regulates the yield of the products, thus providing a transducing measurement for field detection.

6.7.2
Radical Pair Recombination

A radical is a molecule with an unpaired electron spin. By and large, radicals are ionic and have one extra or a missing electron, with $S = 1/2$. When classified

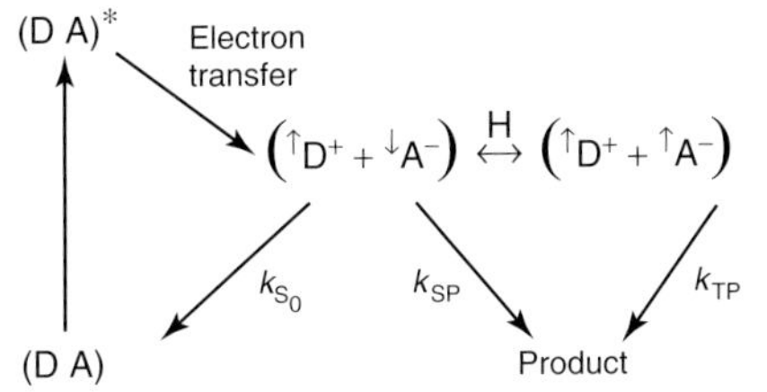

Figure 6.16 Schematic representation of a possible mechanism of radical pair reaction sensitive to the magnetic field. The magnetic field can enhance or reduce the ISC crossing, thus affecting the final product yield. (After T. Ritz in *Procedia Chemistry* **00** (2010), Ref. [3].)

according to their spin multiplicity a radical is a doublet. In the organic solid-state, radicals are often called *polarons*.

When two radicals with spin $S = 1/2$ coalesce into a pair, the total spin can be $S = 1$ or $S = 0$, with triplet and singlet degeneracy. Simple statistics suggest a probability of 75% to form a triplet state and 25% to form a singlet. The most evident proof of the radical pair mechanism is electroluminescence. Opposite charged particles (radicals) injected from the electrodes migrate inside the device and eventually recombine, forming singlet states and triplet states according to the statistics above.

It is important to distinguish between geminate and nongeminate radical pairs. A geminate pair is formed by direct photoactivation or on photoexcitation of a singlet excited state that subsequently breaks into two opposite charges. In geminate radical pairs, the negative charge is bound, or at least correlated to the relative positive charge left behind. On the contrary, when two free radicals meet forming a pair, they are nongeminate. It is the nongeminate pair that follows the statistics presented above, because one spin can be up or down with respect to the other. A geminate pair on the contrary will be in the same spin state as its parent state. Typically, the geminate radical pairs formed on photoexcitation are in the singlet state. These pairs may, however, separate, completely lose their correlation, and later meet with other radicals, equally escaping geminate recombination. This bimolecular process will lead to singlet and triplet states according to statistics. Radical pair recombination in organic solids takes place long time after photoexcitation. Regenerated singlet pairs may afterwards decay radiatively, giving rise to delayed luminescence. Alternatively, emission from the initially excited singlet is called *prompt luminescence*. According to this picture, delayed and prompt luminescence have the same spectral lineshape but very different dynamics. In a simple experiment using a modulated light excitation and lock-in detection, prompt luminescence (PL_F) is in phase with excitation, while delayed luminescence is in quadrature (out of phase) (PL_Q). The quadrature phase is due to the long response time of the delayed luminescence signal. The ratio $PL_Q/PL_F = \omega\tau$, where ω is the excitation modulation frequency, allows estimating the lifetime of the delayed luminescence. Formation of triplet states as a consequence of radical pair recombination can be observed by detecting triplet–triplet absorption or phosphorescence.

6.7.3 Singlet Fission

A singlet excited state, under certain circumstances, may undergo fission into two triplet states. If these states are separated enough, they will show triplet–triplet absorption as individual triplet states. This is thus a mechanism leading to triplet generation even if the overall state remains a singlet ($S = 0$), for the two triplet spins are correlated. SF is a very interesting phenomenon that is receiving attention because of its implication in photovoltaic conversion. The breaking of a singlet state into two triplets suggests a molecular mechanism of carrier multiplication; one photon breaks into two states that can give rise to four charge carriers. However, this process can only yield a true advantage in energy conversion if it takes place from hot states. When this happens, SF is an alternative decay path to heat dissipation, and thus a true gain for photovoltaic conversion. When SF occurs from the lowest excited state, already relaxed, there is no neat gain in energy from splitting of the molecular state. Unfortunately, SF from hot states is not efficient, because of the competition with other internal conversion paths. Concluding, SF may not help photovoltaic conversion, except when a low band gap acceptor is used. In the latter case, high-energy molecular states might not efficiently interact with the acceptor. A dye sensitizer that undergoes fission may thus be needed to convert high-energy photons.

Beside this, SF is a striking evidence of quantum mechanics at work, and for this reason deserves our attention. First we list the experimental observables needed to assign SF.

1) Triplet generation is ultrafast, typically < 1 ps.
2) Triplet formation kinetics does not have correlation with singlet decay
3) Triplet states formed by SF are short lived (0.1–1 ns)
4) Energy conservation requires that the initial excitation energy $E_x > 2E_T$, where E_T is the lowest triplet state energy.

You may compare these requirements with those characterizing spin flip.

1) Triplet formation kinetics matches singlet decay;
2) Triplets are long lived (microseconds to milliseconds)
3) The initial excitation energy $E_x > E_G$, where E_G is the molecular optical gap, usually the energy of the lowest dipole allowed singlet state. (This statement simply assures optical absorption.)

In a pump probe experiment, SF can be easily assigned if triplet–triplet absorption is known. An example is given in Figure 6.17. Following excitation, the transient transmission spectrum clearly shows five different contributions, in different spectral regions. From blue to red spectral region these are photobleaching

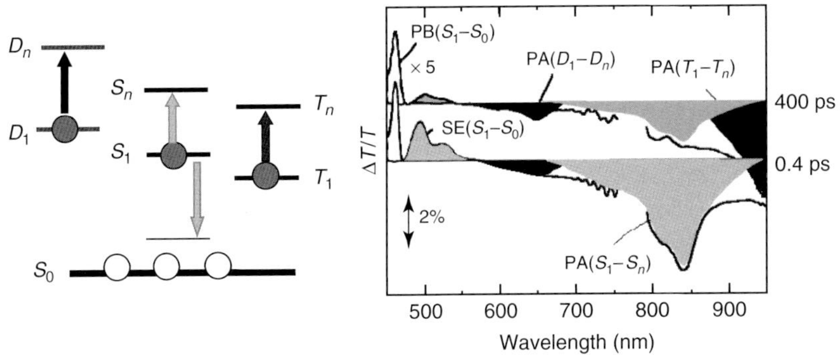

Figure 6.17 The transient transmission spectra of mLPPP polymer film at room temperature after excitation at 390 nm. The color code suggests assignment to singlet, doublet, and triplet. The instantaneous appearance of triplet–triplet absorption at 950 nm indicates that singlet fission is taking place.

due to ground-state depletion; stimulated emission due to $S_1(0)-S_0(1)$ transition (in bracket the vibrational quantum number); doublet (polaron)-induced absorption; singlet-induced absorption; and triplet-induced absorption.

At 400 ps pump probe delay, the presence of the triplet absorption is evident, because of its longer lifetime with respect to the singlet state. However, you can see that triplet–triplet absorption is there from the very beginning, that is, it was formed during pump excitation and it does not have any correlation with singlet decay. This is quite convincing evidence for SF.

If the SF state is above the lowest excited singlet, SF occurs in competition with IC and has a limited yield. If the SF state is the lowest singlet then the material is nonemitting (Kasha rule), while the yield of triplets through SF can be 100%.

The discussion on the nature of the state generating single fission concerns quantum chemistry and the way theory describes excited states. For this reason, it is out of the scope of this book, and we will not go into more details. We just comment that description requires advanced modeling beyond single-particle approach and including electron correlation. The fissioning state is usually described as the combination of two triplet states, correlated to have an overall single character. One interesting thing regards the space extension of the fissioning state in order to observe triplet formation. If the original fissioning state is localized on a single molecule, the property of being described as double excitation (for instance, a pair of interacting triplets) cannot lead to observed triplet separation. The system cannot accommodate two distinguished triplets. On the contrary, when the excited state spans over several molecules (exciton) or extends over many sites of a molecular aggregate or solid, the initial double state can indeed break down into its component, giving rise to a true separation. The initially delocalized wavefunction describing the correlated state will collapse into two localized states separated enough to be identified. For this reason, molecular crystals such as pentacene or long conjugated chains such polydiacetylenes, which support excitonic states, are best candidates to show SF.

Appendix 6.A: Derivation of the Strickler–Berg Relation

From black body theory the radiation density distribution function (J s m^{-3}) per unit frequency is

$$\rho(\nu) = \frac{8\pi h\nu^3 n^3}{c^3}\left[e^{\frac{h\nu}{kT}} - 1\right]^{-1} \tag{6.A.1}$$

At equilibrium

$$W_{la\to ub} = W_{ub\to la} \tag{6.A.2}$$

$$\frac{A_{ub\to la}}{B_{ub\to la}} = \left[\frac{N_{la}}{N_{ub}} - 1\right]\rho(\nu_{ub\to la}) \tag{6.A.3}$$

$$\frac{N_{la}}{N_{ub}} = e^{-\frac{h\nu_{ub\to la}}{kT}} \tag{6.A.4}$$

This leads to Eq. (6.4) $A_{ub\to la} = 8\pi h N_{ub}\nu^3_{ub\to la}n^3c^{-3}B_{ub\to la}$. Considering the whole vibronic manifold and that emission occurs only from the lowest vibrational state "$u0$" we can write $A_{u0\to l} = \sum_a A_{u0\to la}$ and then using Eq. (6.4).

$$\begin{aligned} A_{u0\to l} &= 8\pi hn^3c^{-3}\sum_a \nu^3_{u0\to la}B_{u0\to la} \\ &= 8\pi hn^3c^{-3}K\left|\mu^0_{lu}\right|^2\sum_a \nu^3_{u0\to la}\left|\langle la|u0\rangle\right|^2 \end{aligned} \tag{6.A.5}$$

where the dipole moment has been expanded according to the Condon approximation by introducing the F–C integrals.

Using $\sum_a \left|\langle la|u0\rangle\right|^2 = 1$ we can rewrite

$$\sum_a \nu^3_{u0\to la}\left|\langle la|u0\rangle\right|^2 = \frac{\sum_a \nu^3_{u0\to la}\left|\langle la|u0\rangle\right|^2}{\sum_a \left|\langle la|u0\rangle\right|^2} = \frac{\int \phi(\nu)d\nu}{\int \frac{\phi(\nu)}{\nu^3}d\nu} = \langle \nu_F^{-3}\rangle^{-1} \tag{6.A.6}$$

which allows neglecting constant factors and introducing the experimental PL spectrum because $\int \phi(\nu)d\nu \propto \sum_a \nu^3_{u0\to la}\left|\langle la|u0\rangle\right|^2$. Now we have a relationship between the A coefficient and the fluorescence spectrum. We need now a relationship between the B coefficient and the absorption spectrum. To obtain it we start with the transition rate:

$$W_{l0-ub} = \frac{dN_{l0}}{dt} = N_{l0}B_{l0\to ub}\rho(\nu_{l0\to ub}) = F\sigma N_{l0} \tag{6.A.7}$$

F is the photon fluence (ph/cm^2 s), given by $F = \frac{\rho c}{nh\nu}\mathrm{d}\nu$, so that $B_{l0\to ub} = \frac{c\sigma(\nu_{l0\to ub})}{nh\nu_{l0\to ub}}\mathrm{d}\nu$ and

$$B_{l0\to ub} = \frac{c}{nh}\frac{\sigma(\nu_{l0\to ub})}{\nu_{l0\to ub}}\mathrm{d}\nu = \frac{2303c}{nhN_A}\frac{\varepsilon(\nu_{l0\to ub})}{\nu_{l0\to ub}}\mathrm{d}\nu \tag{6.A.8}$$

Eventually Eq. (6.A.8) could be integrated on the frequency range of the transition centered on $\nu_{l0\to ub}$.

Equation (6.A.8) can also be summed up on all final vibrational states to obtain

$$B_{l0\to u} = \sum_b B_{l0\to ub} = \frac{c}{nh}\int_{l0\to ub}\frac{\sigma(\nu)}{\nu}\mathrm{d}\nu = \frac{2303c}{nhN_{\mathrm{A}}}\int_{l0\to ub}\frac{\varepsilon(\nu)}{\nu}\mathrm{d}\nu \tag{6.A.9}$$

where now the integral is on the whole vibronic band. This expression allows substituting for the dipole moment in Eq. (6.A.5), because $B_{l0\to u} = K\left|\mu^0_{lu}\right|^2$. In doing this we assume that $\left|\mu^0_{lu}\right|^2 = \left|\mu^0_{ul}\right|^2$ and $B_{l0\to ub} = B_{u0\to la}$ and we obtain

$$\frac{1}{\tau_{\mathrm{R}}} = A_{u0\to l} = \frac{8\pi hn_{\mathrm{F}}^3}{c^3}\left\langle\overline{\nu}_{\mathrm{F}}^{-3}\right\rangle^{-1}\frac{2303c}{n_{\mathrm{A}}hN_{\mathrm{A}}}\int_{l0\to ub}\frac{\varepsilon(\nu)}{\nu}\mathrm{d}\nu \tag{6.A.10}$$

which after some rearrangement becomes Eq. (6.1), $\frac{1}{\tau_{\mathrm{R}}} = 2.88\times10^{-9}\frac{n_{\mathrm{F}}^3}{n_{\mathrm{A}}}\left\langle\overline{\nu}_{\mathrm{F}}^{-3}\right\rangle^{-1}\int\varepsilon(\overline{\nu})\mathrm{d}(\ln\overline{\nu})$.

References

1. Domcke, W, Yarkony, D.R., and Koppel, H. (eds) (2000) *Conical Intersections Electronic Structure, Dynamics and Spectroscopy*, World Scientific Publishing Co. Pte., Ltd, New Jersey. ISBN: 981-238-672-6.
2. Englman, R. and Jortner, J. (1970) *Mol. Phys.*, **18**, 145.
3. Ritz, T. (2011) *Procedia Chem.*, **3**, 262–275.
4. Tully, J. (1990) *J. Chem. Phys.*, **93**, 1061–1071.
5. Piryatinski, A. Stepanov, M. Tretiak, S. and Chernyak, V. (2005) *Phys. Rev. Lett.*, **70**, 223001–223005.

7
Vibrational Spectroscopy

7.1
The Semiclassical Picture of the Interaction of Light with Molecules

We live in time–space, but many phenomena regarding electrons and atoms are described in the Fourier conjugated space of frequency and wavevector. There is a way, to see those phenomena in time–space too, which can be instructive and sometimes even useful for quantitative description. Typically, this is true for coherent phonons in molecules or solids.

Let us consider the phononless electronic response of a single oscillator (or two-level system).

The linear polarization is linked to the electric field by the susceptibility, $P^{(1)}(\omega) = \varepsilon_0 \chi(\omega) E(\omega)$, where $\chi(\omega) = \omega_{\mathrm{P}}^2 \left(\omega_{\mathrm{eg}}^2 - \omega^2 - i2\gamma\omega\right)^{-1}$ is frequency dependent because the response is in general not instantaneous. By Fourier transform $\chi(\omega)$, one can obtain the response function in the time domain as $K_{\mathrm{g}}(t) = \int_{-\infty}^{+\infty} \chi(\omega) \mathrm{e}^{-i\omega t} \mathrm{d}\omega = \omega_{\mathrm{p}}^2 \mathrm{e}^{-\gamma t} \frac{sen \nu_{\mathrm{eg}} t}{\nu_{\mathrm{eg}}} H(t)$, where $H(t)$ is the Heaviside function. The electronic "correlator" $K_{\mathrm{g}}(t)$ is a damped oscillator that describes the polarization decay under impulsive excitation. The damping constant, "γ," is the electronic dephasing, and ν_{eg} is the transition electronic frequency. The $H(t)$ function imposes causality. The corresponding expression for polarization in time domain is the convolution of the electric field with the time-dependent response function, $\mathrm{P}^{(1)}(t) = \left(\mathrm{i}/\hbar\right) E \otimes K_{\mathrm{g}}(t)$. The electronic transition lineshape in frequency is the Fourier transform of the response function in time. So we have two pictures, in time and frequency. The process of polarization decay in time leads to the Lorentzian lineshape in frequency.

Now we can consider the lineshape of the vibronic transition. This brings about the concept of wavepackets in molecular dynamics. Whether this picture bears more truth than the standard Franck–Condon picture is left as an open philosophical question. The semiclassical picture however provides some insight on what happens to the nuclei after the electronic transition took place, which is missing in the frequency domain explanation based on steady-state wavefunction overlap.

The Photophysics behind Photovoltaics and Photonics, First Edition. Guglielmo Lanzani.

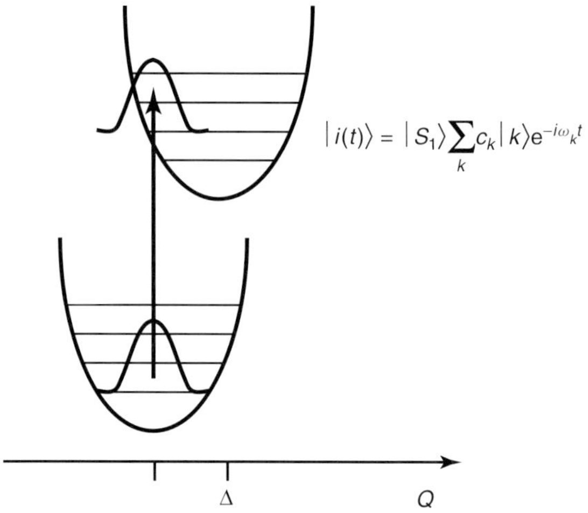

Figure 7.1 Coherent superposition of vibrational states.

After the sudden rearrangement of the electronic cloud due to the electronic transition, the slow nuclei are out of equilibrium, subject to a restoring force. The Franck–Condon idea that nuclei could not move during electronic transition means in quantum mechanics that their wavefunction is unchanged during electronic transition. If electron–phonon coupling is not zero, this wavefunction is no more an eigenstate of the system, and it can be represented only as a superposition of excited-state vibrational wavefunctions. This means that the vibrational state is now a wavepacket, as depicted in Figure 7.1. The wavepacket will slide downhill, following at first the path of the steepest descent, changing shape. This path very closely approximates a classical path, at least for a few vibrational periods. The expectation values of position and momentum follow the classical trajectory of motion, and the wavepacket oscillates back and forth "on" the upper state surface.

Half a century ago, well before ultrafast lasers were developed, spectroscopists knew that all this could be "physically" achieved by using very short light pulses, as W. T. Simpson, and D. L. Peterson state in their 1957 article: "This freezing-in of the ground-state wavefunction could be accomplished in practice by using a sufficiently short pulse of light . . ." [1].

It turns out that this wavepacket motion is strictly correlated to the vibronic absorption spectrum. According to Eric J. Heller, "*The absorption lineshape is the Fourier Transform of the overlap of the wavepacket moving on the excited electronic surface* $|i(t)\rangle$ *with itself at time zero* $|i(0)\rangle$, in formula

$$\alpha(\omega) = C\omega\left\{\mathrm{Re}\int_0^\infty \mathrm{d}t\,\langle i(0)|i(t)\rangle \mathrm{e}^{i(\omega-\omega_{\mathrm{eg}})t-\gamma t}\right\} \tag{7.1}$$

where C is a constant and γ is a phenomenological damping constant that we can assign to the electronic dephasing. Now let us first look at Eq. (7.1), let alone where it comes from. It says that the spectrum at frequency ω is the Fourier transform of the dynamics following an instantaneous Franck–Condon transition at $t = 0$. Once

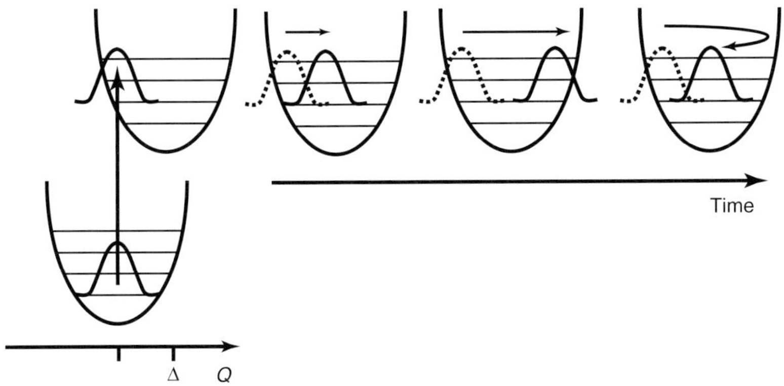

Figure 7.2 Evolution in time of the wavepacket on the upper state surface.

in the excited electronic state, $i(t)$ is a nonstationary state that evolves according to the time-dependent Schrödinger equation $\left|i(t)\right\rangle = e^{-iH_e t/\hbar}\left|i(0)\right\rangle$, where H_e is the vibrational Hamiltonian of the upper surface. In the approximation of displaced oscillators, H_e is the harmonic Hamiltonian with displaced equilibrium position with respect to the ground state. The correlation function in Eq. (7.1) evaluates the overlap between $i(t)$ and the original $i(0)$ at the birthplace (Figure 7.2). Making this dynamics explicit, the time-dependent response function is

$$K_g(t) = e^{-\gamma\tau} e^{-i\omega_{eg}t}\left\langle g\left|\mu^*_{eg} e^{-iH_e t/\hbar}\mu_{eg}\right|g\right\rangle = e^{-\gamma\tau} e^{-i\omega_{eg}t}\left\langle i(0)\middle|i(t)\right\rangle \tag{7.2}$$

and the absorption lineshape is $\Phi(\omega) = i\int_0^\infty ds e^{i\omega s} K_g(s)$, that is, as before the real part of Fourier transform of $K_g(t)$.

If we picture the wavepacket dynamics and its correlation we can understand all the features as appearing in the vibronic spectrum in frequency. Figure 7.3a shows $e^{i\omega_{eg}} K_g(t)$ (i.e., the overlap without the optical frequency). On the basis of the uncertainty principle,[1] at each feature in time we can associate a feature in frequency. At $t = 0$ the correlation is maximum. Then $\left\langle i(0)\middle|i(t)\right\rangle$ starts to decay because of $i(t)$ departure. This is the fastest process, and according to the uncertainty principle it corresponds to the broadest feature in the spectrum: its width. The steeper the potential the faster $i(t)$ goes away and the broader the vibronic envelope. This happens if the vertical transition reaches a well out-of-equilibrium position, when there is strong electron–phonon coupling. If T_A describes the initial decay of the correlation function, $1/T_A$ is the spread of the vibronic envelope in frequency.

1) The Heisemberg uncertainty principle regards conjugated operators and their observables. Because time and frequency are not operators in quantum mechanics, the relationship between Δt and $\Delta\nu$ is not an uncertainty relationship in the strict Heisemberg definition, but it describes a limitation in measuring.

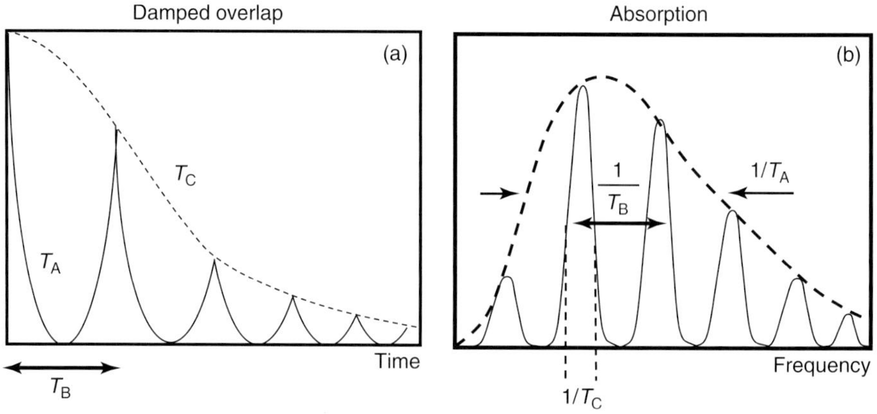

Figure 7.3 Damped wavepacket correlation (a), and its Fourier transform (b). The features in time and frequency are specified. (This picture is reproduced from Figure 3 of Ref. [2]).

After the initial fall off, $\langle i(0)|i(t)\rangle$ stays low until $t \approx T_B$, when the wavepacket returns to its birthplace (vibrational recurrence). T_B is the vibrational period of the wavepacket oscillation. The periodic round-trip of the wavepacket provides vibrational replicas in frequency, equally spaced by $1/T_B$. Since $i(T_B)$ is displaced and spread relative to $i(0)$, because of many perturbations, $\langle i(0)|i(T_B)\rangle < \langle i(0)|i(0)\rangle$ is always true. In the ideal model with harmonic potential surface, the wavepacket would not suffer any change during a cycle; however, electronic dephasing acts on it. Ultimately the amplitude of the overlap will be zero, and the time scale of this decay (T_C) gives the width of each peak in the spectrum as shown in Figure 7.3b.

The vibrational recurrence is a physical observable in time domain, provided that light pulses short enough are used in a pump probe experiment. The real observable is, however, not a single molecule but an ensemble of molecules, and recurrence, as periodic oscillations, can be observed as long as ensemble coherence is preserved. In other words, each molecule in a semiclassical state can be represented by a pendulum. What can be measured is the collective oscillation of a set of pendula. Loss of this collective coherence will result in deletion of the signal, in spite of a perseverance of the coherent state in each molecule (Figure 7.4).

What happens when cw, monochromatic light hits the sample, and which is the relationship with ultrafast dynamics? When we turn on the light at frequency ω, "little pieces" of the ground-state wavefunction (remember probability is what quantum mechanics is all about) are constantly being brought up to the excited-state surface, with phase $e^{i\omega\tau}$, where they propagate. When a new piece arrives, it finds others already there, described by $i(t)$ at various times t. As these pieces return to their birthplace, new pieces come up, giving rise to constructive or destructive interference, causing absorption intensity to depend on ω. The interference of many events will ultimately leave one single state, spread over the molecular space, which represents the eigenstate with resonant energy ω, so that the standard

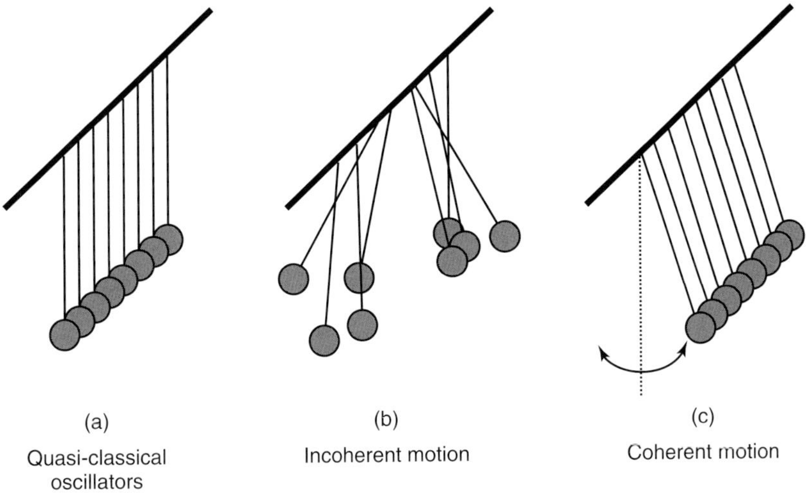

Figure 7.4 In (a) each molecule in the quasi-classical state is represented by a pendulum. (b) During incoherent motion the average position is fixed. (c) The observable is the collective motion, due to the coherent state: all pendula behave like a single macroscopic oscillator.

notion on Franck–Condon overlap is restored. The long observation time, required to define ω essentially washes out the periodic dynamics, which is, however, the fundamental event behind the absorption value at that particular frequency ω.

7.2 Derivation of the Correlation Function

We show here a simplified derivation of the correlation function for the vibronic absorption lineshape, which disregards the effect of temperature. Let us start with the sum-over-states expression for the absorption spectrum without line broadening:

$$\alpha(\omega) = C \sum_{n} |\langle n|i\rangle|^2 \delta(\omega - \omega_n) \tag{7.3}$$

where C is a set of constants. We now expand the modulus and use the integral definition for the delta function:

$$\alpha(\omega) = \int_{-\infty}^{\infty} \mathrm{d}t \sum_{n} \langle i|n\rangle \langle n|i\rangle \exp\left[i(\omega - \omega_n)t\right] \tag{7.4}$$

Then we bring $\mathrm{e}^{-i\omega_n t}$ into the bracket product, and use the definition for the expectation value:

$$\alpha(\omega) = \int_{-\infty}^{\infty} \mathrm{d}t \sum_{n} \langle i \mid n\rangle \left\langle n \mid \exp\left[-iH_{\mathrm{ex}}t\right]/\hbar \left|i\right.\right\rangle \exp(i\omega t) \tag{7.5}$$

Now what we have in the second product is the time evolved state of the initial state, $i(t)$, and this leads us to the expression

$$\alpha(\omega) = \int_{-\infty}^{\infty} \mathrm{d}t \sum_n \langle i|n\rangle \langle n|i(t)\rangle \exp(i\omega t) \tag{7.6}$$

Adopting the "closure" rule for the complete orthonormal system, and adding a phenomenological damping, we get the final result:

$$\alpha(\omega) \propto \mathrm{Re} \int_0^{\infty} \mathrm{d}t \langle i(0)|i(t)\rangle \mathrm{e}^{i\omega t - \gamma t} \tag{7.7}$$

If the wavepacket is described by a Gaussian function, its propagation can be done analytically, an easy task at least for one vibrational coordinate. The wavepacket propagation is

$$|i(t)\rangle = \mathrm{e}^{-iH_\mathrm{e}t/\hbar} |i_0\rangle = \left(\tfrac{1}{\pi}\right)^{\frac{1}{4}} \exp\left[-\alpha_t (q - q_t)^2 + ip_t (q - q_t) + i\gamma_t\right] \tag{7.8}$$

and the overlap

$$\langle i_0|i(t)\rangle = \left(\frac{1}{a}\right)^{1/2} \exp\left(-(p')^2/4a + i\gamma'\right) \tag{7.9}$$

with the following parameters and functions:

$$\begin{aligned} a &= \alpha_t + \frac{1}{2} \\ p' &= p_t - 2i\alpha_t q_t \\ \gamma' &= i\alpha_t q_t^2 - p_t q_t + \gamma_t \end{aligned} \tag{7.10}$$

$$\begin{aligned} \alpha_t &= \frac{1}{2} \\ p_t &= \Delta \sin \omega t \\ q_t &= \Delta (1 - \cos \omega t) \\ \gamma_t &= \frac{-\omega t}{2} - \frac{\Delta^2}{2} \sin \omega t \cos \omega t - \Omega_0 t/\hbar \end{aligned}$$

These simple equations, introduced into Eq. (7.2) allows evaluation of the absorption lineshape.

7.3 The Full Vibronic Correlator in Time

The cross section of absorption is $\sigma_\mathrm{A} = \frac{4\pi\mu^2\omega}{3\hbar c} i \int_0^{\infty} \mathrm{d}s\, \mathrm{e}^{i\omega s} K_\mathrm{g}$, where the full multimode, temperature-dependent correlator is $K_\mathrm{g} = \exp(-i\Omega_{00}s - \Gamma s - g(s))$. Here the new term is the vibrational function, $g(s)$, which is a sum over the vibrational modes:

$$g(s) = \sum_k g_k(s) = \sum_k \tfrac{1}{2}\Delta_k^2 \left[(\overline{n}_k + 1)\left(1 - \exp(-i\omega_k s)\right) + \overline{n}\left(1 - \exp(i\omega_k s)\right)\right] \tag{7.11}$$

Each vibrational mode is specified by Δ_k, n_k, and ω_k, the dimensionless displacement, the population, and the frequency, respectively. The phonon population term that bears the temperature dependence is

$$\bar{n}_k = \left(\exp\left(\frac{\hbar\omega_k}{k_B T}\right) - 1\right)^{-1} \quad (7.12)$$

Having in the sum in Eq. (7.11) a small frequency vibrational mode (<10 cm^{-1} in wavenumber) with very large electron–phonon coupling ($\Delta \gg 1$) provides a Gaussian-like broadening mimicking solvent/environment effect. Without this trick, the electronic dephasing Γ will provide a Lorentzian lineshape to the spectral features. The formula above is easily evaluated numerically, and for many vibrational modes this formalism is definitively easier than the full Frank–Condon expression in frequency domain.

7.4 Raman Scattering

An electromagnetic wave impinging and traversing a medium generates a polarization, $P(t)$. This polarization acts as a source for a secondary wave, which interferes with the original wave and gives rise to a transmitted wave. If absorption does not take place, amplitude is preserved and the response function linking polarization and field is real. Only a phase shift between the incoming and outgoing wave is obtained on traveling through the medium. The detailed mechanism of secondary wave generation and interference is complex. Essentially one can think that the incoming wavefront breaks in point sources of secondary spherical waves. This is highly idealized because a point source does not exist, yet it allows depicting the phenomenon as a wave reconstruction according to the Huygens model. The result is that in a perfect medium light travels in straight path. However, any perturbation in the homogeneous medium can cause scattering. Scattered light is distributed all over the solid angle, and may contain frequencies different from the incident light (Figure 7.5). Let us assume that monochromatic light is propagating through a material. Any imperfection or perturbation, e.g. a transient density or polarizability modulation can induce scattering. The larger fraction of scattered light is at the same frequency of the incident one, a phenomenon known as *Rayleigh scattering*. A small part of the scattered light contains characteristic frequencies that are shifted from the incident frequency by rotational or vibrational quanta. This is called *Raman scattering*, and it is of fundamental importance in material characterization. Raman bands are characteristic of chemical bonds, depending on the local geometry in a well-known way, thus being the fingerprint of the molecular structure. Experimentally, the scattered light can be collected in many different geometries, all designed to remove the much stronger (10^7 times) incident radiation (Figure 7.6). A Raman spectrum shows a series of peaks at frequencies that are usually measured with respect to the incident frequency (this frequency shift is called *Raman shift*). Depending on the technique and properties of the

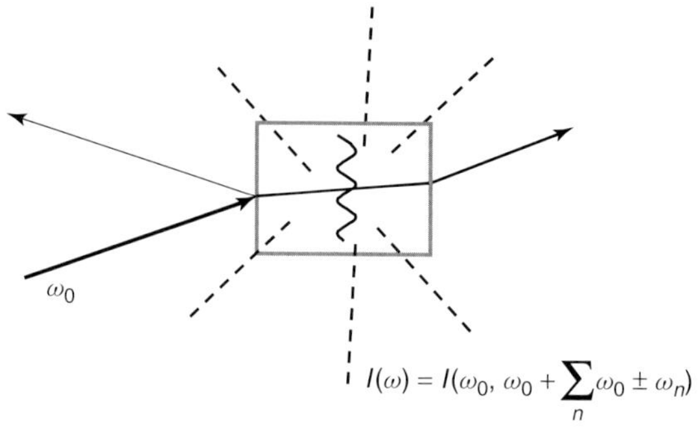

Figure 7.5 Transmitted, reflected, and scattered light.

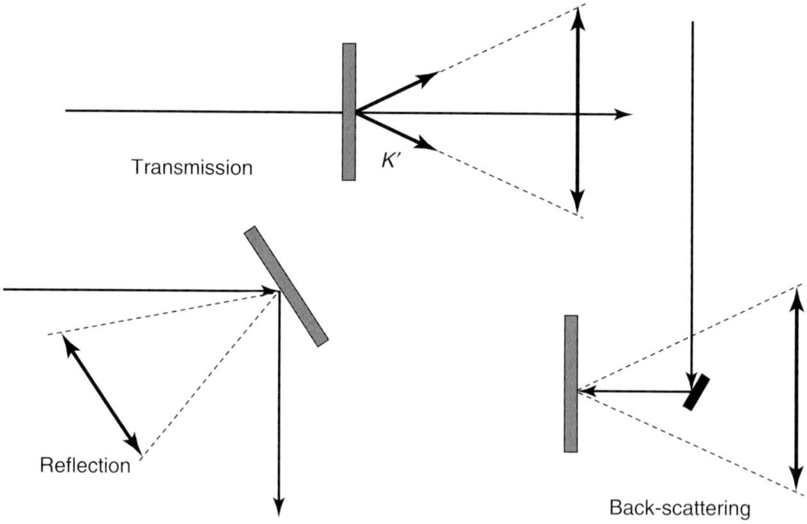

Figure 7.6 Three different experimental geometries for Raman scattering. The double-headed black arrow is a collecting optics.

experimental setup, one can look at Raman shifts spanning from tens of wavenumbers to thousands. In most cases, however, it is difficult to measure Raman shifts smaller than $100\,\text{cm}^{-1}$, because of the strong excitation and Rayleigh peak.

Classical physics provides a fairly simple and intuitive explanation for Raman scattering, according to the Placzek model stated in 1934. Let us assume that molecular polarizability depends on the vibrational coordinate Q, giving $p = \varepsilon_0\alpha(Q(t))E(t)$. Expanding α in terms of the local coordinate $Q(t)$ to the first order gives $\alpha(Q) = \alpha_0 + \left(\frac{\partial\alpha}{\partial Q}\right)Q$. We assume now that the vibrational motion is harmonic, $Q(t) = \tilde{Q}\cos(\omega_m t)$. The molecular dipole moment induced by an external, monochromatic electric field at frequency ω_0 will be $p = \varepsilon_0\left[\alpha_0 + \alpha'\tilde{Q}\cos(\omega_m t)\right]\tilde{E}\cos(\omega_0 t)$. This product leads to $p = \varepsilon_0\alpha_0\tilde{E}\cos(\omega_0 t) + \varepsilon_0\alpha'\tilde{Q}\tilde{E}\cos(\omega_0 \pm \omega_m)t$, which describes a time-dependent dipole moment (eventually a medium polarization). The first term oscillates at the same frequency of the monochromatic excitation and contributes to Rayleigh scattering. The second term oscillates at frequency $(\omega_0 - \omega_m)$ and $(\omega_0 + \omega_m)$ and accounts for the Raman effect, Stokes scattering, and anti-Stokes scattering, respectively. The classical picture, however, cannot explain a few important experimental observations:

1) By tuning the excitation frequency, some Raman peaks shoot up in amplitude, dominating the spectrum (a phenomenon known as resonant Raman scattering);
2) Anti-Stokes Raman scattering is much weaker than the Stokes one and shows a peculiar temperature dependence;
3) Not all vibrations are Raman active; very few are resonant, and when this happens, series of overtones and combination bands can be seen, with large intensity.

Quantum mechanics describes Raman scattering as a two-photon process. In a Stokes event, a photon at energy $\hbar\omega_0$ is destroyed and a photon at energy $\hbar\omega_S$ is created, together with a material phonon $\hbar\omega_m$. Energy conservation implies $\hbar\omega_S = \hbar\omega_0 - \hbar\omega_m$, and in solids momentum conservation requires $k_0 - k_S = k_m \cong 0$. The last equation stems from the observation that photons have wavevectors much smaller than the typical Brouillon zone, so that Raman scattering occurs essentially at $k = 0$.

The rate of spontaneous Raman scattering transition depends on the photon (n_{L}), and the phonon (n_v) populations according to $W_{\mathrm{Stoke}} \propto |\langle n_{\mathrm{L}} - 1, 1, n_v + 1 \,|\, a_{\mathrm{L}} a_S^+ a_v^+ \,|\, n_{\mathrm{L}}, 0, n_v, \rangle|^2 = n_{\mathrm{L}}(n_v + 1)$ where a_{L} destroy one incident photon, a_S^+ creates a Stokes photon, and a_v^+ creates a phonon. As a result, the Raman Stokes rate is linear with the incident photon population (intensity) and depends on the phonon population *increased* by one. This assures that the probability of the Raman Stokes scattering remains sizable even when the phonon population is negligibly small. Not so for the anti-Stokes scattering, whose rate $W_{\text{anti-Stokes}} \propto |\langle n_{\mathrm{L}} - 1, 1, n_v - 1 \,|\, a_{\mathrm{L}} a_{\mathrm{AS}}^+ a_v \,|\, n_{\mathrm{L}}, 0, n_v \rangle|^2 = n_{\mathrm{L}} n_v$ is proportional to the phonon population. This can be very small for phonon energy $\hbar\omega_m > K_{\mathrm{B}} T$. Stokes and anti-Stokes spontaneous Raman scattering transitions are depicted in Figure 7.7. Quantum mechanics provides a simple explanation for the experimental observation of much weaker anti-Stokes Raman bands. The ratio between Stokes and anti-Stokes Raman amplitudes for a known vibration, $\frac{(n_v+1)}{n_v}$, depends on phonon population, and it can be used to estimate the sample temperature. Eventually a micro-Raman probe could provide information on local temperature.

By using second-order perturbation theory, the Raman scattering cross section is described by a sum-over-states expression with factors $\left[\left(E_{en} - E_{gi} - \hbar\omega_0\right) - i\Gamma\right]^{-1}$ weighting each state contribution. When the excitation energy approaches a vibronic gap $E_{en} - E_{gi} \approx \hbar\omega_0$, the sum collapses into a single term for the electronic transition $g \rightarrow e$. This explains the resonant amplitude increase in the Raman spectrum. Within the adiabatic approximation and the Condon approximation, the

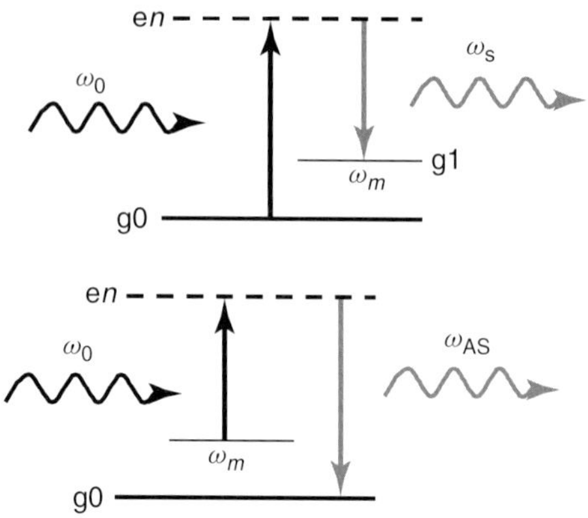

Figure 7.7 Stokes and anti-Stokes Raman transitions.

Franck–Condon resonant Raman scattering cross section is given by

$$\sigma_R = \frac{2\mu_{eg}^4 \omega_0 \omega_S^3}{9\varepsilon_0 c^4} \sum_i B_i \left| \sum_n \frac{\langle f|n\rangle \langle n \mid i\rangle}{\left(E_{en} - E_{gi} - \hbar\omega_0\right) - i\Gamma} \right|^2 + \mathrm{NR} \tag{7.13}$$

where NR is a nonresonant contribution. Here "μ" is the transition dipole moment of the electronic transition $gi\rangle \rightarrow en\rangle$. $n\rangle$ is the vibrational wavefunction of the intermediate electronic state, $i\rangle$ and $f\rangle$ the vibrational wavefunctions of the initial and final states, both in the ground electronic state. The operator B_i specifies the thermal population. Usually $i = 0$, $f = 1$, and B_i can be neglected. Equation (7.13) shows that overtones can appear in the spectrum (for $f = 2, 3, \ldots$) and bear the similarity with the Franck–Condon absorption term. Again the Franck–Condon overlap integrals define the strength of the transition. In addition, symmetry of the wavefunctions comes into play, and we see that Franck–Condon total symmetric modes that are active in absorption can also be strong in Raman scattering. This explains the selective enhancement of few modes in resonance. On scanning the excitation wavelength while looking at a particular Raman peak amplitude, one obtains the Raman excitation profile. This quantity can be well reproduced also using the time correlator as introduced above for the absorption lineshape. The Raman excitation profile for a vibrational mode "m" is given by

$$\sigma(\omega) \propto \sum_i B_i \left| \int_0^\infty \mathrm{d}t \langle f|i(t)\rangle e^{i(\omega-\omega_0)t-\gamma t} \right|^2 \tag{7.14}$$

where correlation is between the wavepacket $i(t)$ with the first vibrational state of the ground (initial) electronic state, $f\rangle$. The full expression for the Raman cross section is

$$\sigma_R^m = \frac{2\mu^4 \omega_S^3 \omega}{9\hbar^2 \varepsilon_0 c^4} \left| \int_0^\infty \mathrm{d}t\, e^{i\omega t} K_R \right|^2 \tag{7.15}$$

where the lineshape correlator is $K_R = (e^{-i\omega_m t} - 1)K_g$ (Figure 7.8). Using this expression for each Raman active mode, one can easily reproduce the excitation

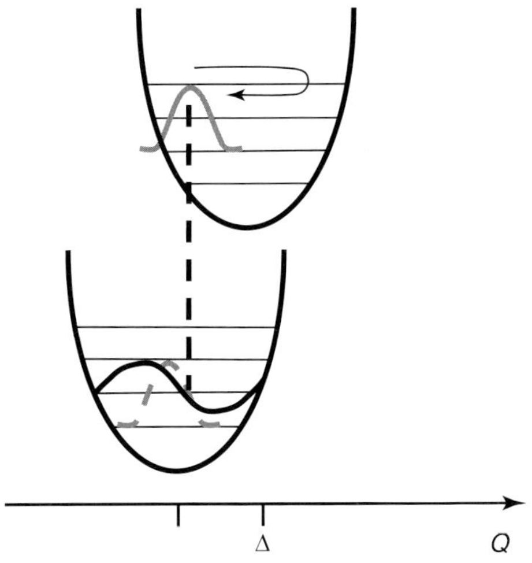

Figure 7.8 Raman cross section in time domain.

Raman profile, or given a monochromatic excitation, reproduce the resonant Raman spectrum $I(\omega_S) = \sum_{m=1}^{N} \sigma_R^m(\omega_0) g(\omega_S^m)$.

The spontaneous Raman lineshape "g" is given by Fourier transform of the correlation function of the vibrational coordinate: $g(\omega) \propto \int_{-\infty}^{\infty} dt \langle Q(0)|Q(t)\rangle_{eq} e^{i(\omega-\omega_0)t}$ (for the homogeneously broaden vibrational system). This gives rise to a Lorentzian, $g(\omega_S^m) = \frac{\gamma}{\left[\omega_S^m - \omega_S\right]^2 + (\gamma)^2}$, where $\gamma = \frac{1}{T_2^v}$ is the vibrational damping constant. The vibrational damping time corresponds to the decay of the coherence in the nuclear displacement $\langle Q\rangle$. This is the result of the fluctuation–dissipation theorem, which states that the vibrational correlation function $\langle Q(0)|Q(t)\rangle_{eq}$ at equilibrium and the coherent amplitude $\langle Q\rangle$ decay exponentially with the same time constant T_2^v. The equations of motion describing vibrational excitation are very similar to the optical Bloch equations, where the collective polarization P is substituted by the collective displacement $\langle Q\rangle$. Displacement and population are linked by the equation of motion:

$$\frac{d^2\langle Q\rangle}{dt^2} + \frac{2}{T_2^v}\frac{d\langle Q\rangle}{dt} + \omega_v^2\langle Q\rangle = \frac{F(z,t)}{m}\left[1 - 2(n+\tilde{n})\right] \tag{7.16a}$$

$$\frac{\partial}{\partial t}n + \frac{1}{T_1^v}n = \frac{1}{\hbar\omega_v}F(t)\frac{\partial}{\partial t}\langle Q\rangle \tag{7.16b}$$

n represents the occupation number in excess of the thermal equilibrium value $\tilde{n}$, m, and ω_v are the reduced mass and frequency of the vibrational mode. F is the effective force exerted by the electromagnetic field onto the vibrating molecule, obtained as gradient of the electromagnetic energy density, $F(z,t) = \frac{1}{2}\varepsilon_0 \left(\frac{\partial\alpha}{\partial Q}\right)_0 \overline{E}\cdot\overline{E}$. Typical of stimulated Raman scattering is the use of an exciting field $\overline{E}(z,t) = \frac{1}{2}\hat{e}_1 E_1(z)e^{i\omega_1 t} + \frac{1}{2}\hat{e}_2 E_2(z)e^{i\omega_2 t} + c.c.$ such that the force is proportional to $\overline{E}\cdot\overline{E} = \frac{1}{4}\hat{e}_1\cdot\hat{e}_2 E_2(z)E_1^*(z)e^{i(\omega_2-\omega_1)t} + c.c.$ If the frequency difference

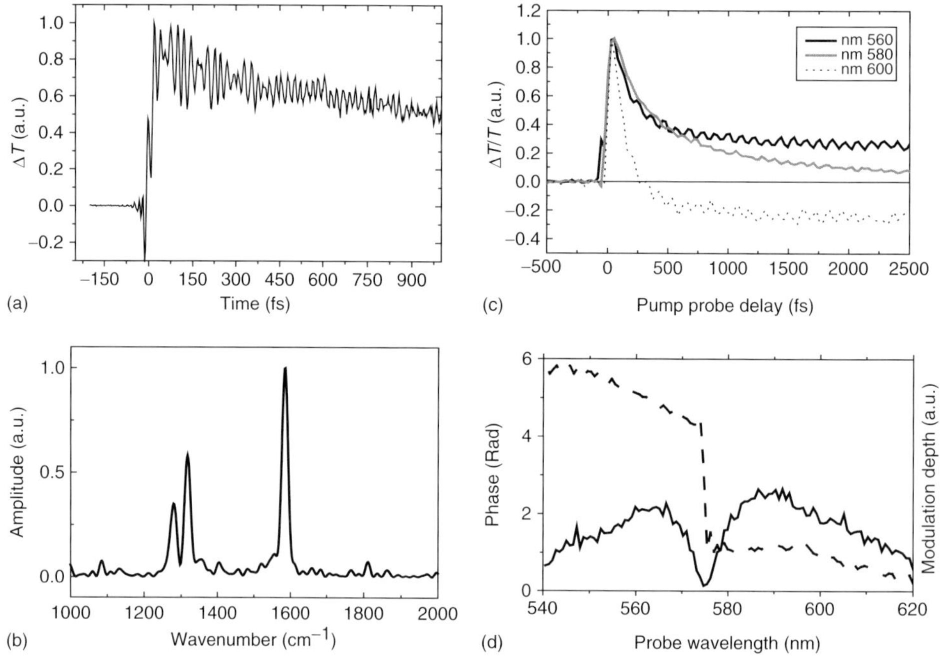

Figure 7.10 Coherent phonons observables. (a) Time trace for PPV at 580 nm probe. (b) The Fourier amplitude spectrum. It shows the frequency components, accounting for the beating in the time trace above, due to multimode oscillation. (c) Several time traces for semiconducting carbon nanotubes of chiral indexes (6,5), modulated by the radial breathing mode at 307 cm^{-1}. (d) The modulation depth profile showing the characteristic minimum and the phase profile, with the 2π shift that both occur at the electronic resonance (here the narrow second exciton peak).

modulation depth profile and the phase dependence on the probing wavelength provide information about the potential energy surface (PES) sampled by the vibrational wavepacket (Figure 7.10).

The observation of vibrational wavepacket motion allows detecting molecular dynamics in real time. Such time domain investigation provides information on ground and excited-state vibrational frequencies, dephasing, anharmonicity, and nonadiabatic coupling. Very often, a coherent phonon time trace contains many frequencies, which can be singled out by Fourier transform or other numerical analysis, yielding amplitudes, phase, and damping time constants of each mode. In general, vibrational coherence is initiated in both the excited and ground states, and the corresponding spectroscopical signatures overlap in a large spectral region. To separate the two is not trivial. They are expected to have different phases, but this is not always easy to measure. Temperature dependence is different, yet the effect is very weak for phonon energy exceeding $k_B T$; in addition, a cryostat introduces dispersive optics, which requires further compensation and worst experimental conditions. Mode assignment based on theory is of little help, given the difficulty in working out reliable excited-state normal modes. A few clear cases, however, exist, wherein assignment can be done quite reasonably:

1) The pulse duration (t_p) is much shorter than the observed vibrational period $\tau_v(\tau_v \approx 10t_p)$. In this case, the δ-like (strictly impulsive) excitation condition is fulfilled, and coherent motion is only initiated in the excited state.
2) Relaxation processes quickly destroy excited-state coherence (e.g., internal conversion, exciton self-trapping, etc.). In this case, the coherent signal detected after relaxation will necessarily contain only ground-state contribution.
3) Excited-state modes are quite different from ground-state ones, and assigned (by other experiments or straightforward calculations). This situation, rarely encountered (see Box 7.2), implies that a simple comparison of the transient modes with those in the conventional Raman spectrum leads to the assignment.

As a remark on the last point, observation of well-known molecular dynamics in real time always comprises additional information to conventional techniques,

Box 7.2: Duschinsky Rotation

Duschinsky rotation occurs when geometry and/or force constants are different between ground and excited state. Excited-state normal coordinates are formed from linear combinations of ground-state coordinates of the same symmetry. Duschinsky rotation has several effects on the vibrational features of the molecule:

1) It tends to increase the cross section for excitation of combinations of ground-state modes in the Raman process;
2) The frequencies of the resulting excited-state modes can be significantly different from the corresponding ground-state frequencies. Mixing can lead to increase/decrease in frequency in the excited state; excited-state frequencies are not linear combinations of ground-state ones, only coordinates are. Frequencies derive from the curvature of the potential surface along the new coordinate;
3) Some wavepacket motion can be induced along a mode with no excited-state displacement if it undergoes Dushincky-type mixing with another mode.

Equation (7.18) specifies how the ground-state normal coordinates Q_g can be transformed into the excited-state normal coordinates Q_e:

$$Q_e = SQ_g + D \tag{7.18}$$

Here S is the rotation matrix characterized by the angle θ_D and D is the vector containing the displacements in the excited-state modes. For a two-mode system S is written as

$$S = \begin{bmatrix} \cos\theta_D & \sin\theta_D \\ -\sin\theta_D & \cos\theta_D \end{bmatrix} \tag{7.19}$$

which represents a rotation in plane.

so that it may be worth even in well-known situations. The clear case here is ground-state coherence. In principle, this is detected by standard, frequency domain Raman scattering, using monochromatic excitation. Yet, impulsive Raman scattering off the ground state have points. Dephasing as calculated from Raman linewidth regards the 0–1 vibrational transition of the active mode. Impulsive Raman scattering contains coherence from many vibrational levels, up to higher quantum numbers. This provides an estimate of the lifetime of higher levels and anharmonicity. Anharmonic coupling between modes can be observed as a modulation in the vibrational parameters. This effect, which is subtle and very hard to detect, may be better resolved in the time domain. Finally, low-frequency phonons, crucial in fields such as protein conformation, are much better and easily detected in the time domain.

There are three generation conditions for coherent phonons: (i) impulsive absorption (IA), (ii) impulsive stimulated resonance Raman (ISRR) scattering, and (iii) impulsive stimulated Raman (ISR) scattering.

1) IA within the Franck–Condon picture, can be described easily by using the time argument (freezing-in of the nuclear motion) (Figure 7.11). A δ-like pulse ($t_P \gg \tau_{vib}$) projects the ground-state vibrational wavefunction onto the excited-state PES. The wavepacket has initially zero momentum and kinetic energy, but it is displaced from equilibrium. The generation process can also be discussed in terms of frequency. The broad-pulse spectrum comprises transitions to many vibrational levels of the excited state, all excited in phase, giving rise to the coherent superposition leading to the nonstationary wavepacket.
2) ISRR scattering regards ground-state coherence and is generated by a Raman-type interaction. In terms of frequency you can think that, within the broad pulse there are excitation and Stokes frequency pairs that force the vibrational oscillation, similar to stimulated Raman scattering. This induces

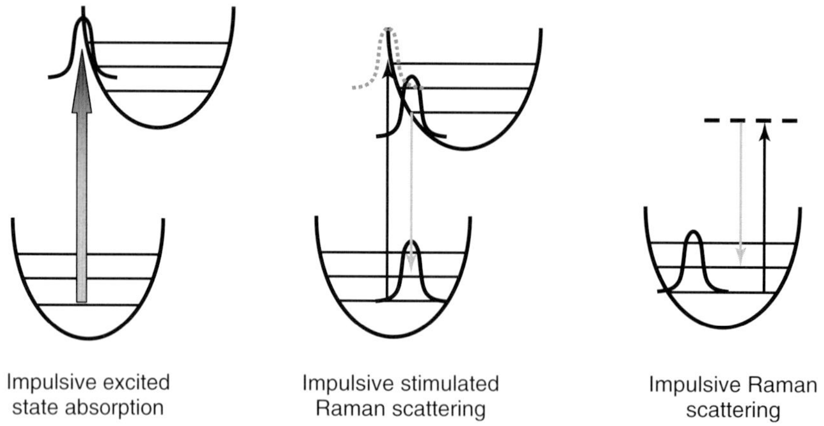

Figure 7.11 Three different excitation processes of a vibrational coherence.

both displacement and momentum of the initial wavepacket in the ground state. Because the ground state is populated, theory suggests that this initial wavepacket is actually a transient "hole" or distortion in the ground-state wavefunction. When in resonance, the transient evolution of the excited state during pulse interaction should be taken into account. The pump pulse places amplitude on the excited-state PES. The nascent wavepacket starts sliding down the potential surface, while the pump pulse is still acting on the system. Further interaction with the pump field couples back wavepacket amplitude to the ground state, forming a ground-state vibrational wavepacket (a projection of the displaced excited-state wavepacket). Because this occurs within the pump pulse, a very short pump pulse with respect to the vibrational period (δ-like excitation) will not lead to a displaced ground state because no motion will take place during the interaction. A very long pump pulse will behave like a cw excitation, so the optimum is in between, with a pump pulse duration in the order of 1/3 to 1/4 of the vibrational period.

3) ISR scattering is the non resonant mechanism. It is better understood in the frequency domain. A very short pulse with respect to vibrational period will contain pairs of excitation and Stokes or anti-Stokes wavelengths. Those will act on the system, forcing its oscillation, similar to a kick to a classical oscillator. The ground-state wavepacket will initially acquire momentum. Figure 7.11 sums up the three excitation conditions.

We now briefly discuss a full quantum mechanical (nonperturbative) approach to the calculation of coherent phonon spectra (amplitude and phase) developed by Kumar *et al.* [3]. In their expression, nonstationary transmission is associated with an effective third-order susceptibility, which is formally a linear susceptibility, but with explicit pump probe delay dependence, $\Delta\chi(t,\tau)$. To start with, let us recollect the link between linear susceptibility and time correlator:

$$\chi^{(1)} = \frac{i\left|\mu_{eg}\right|^2}{\hbar}\left[K_g(t-\tau) - K_g^*(t-\tau)\right] \tag{7.20}$$

$\Delta\chi(t,\tau)$ is associated with the time-dependent correlator in a similar way:

$$\Delta\chi = \frac{i\left|\mu_{eg}\right|^2}{\hbar}\left[C_g(t,\tau) - C_g^*(t,\tau)\right] + \frac{i\left|\mu_{eg}\right|^2}{\hbar}\left[C_e(t,\tau) - C_e^*(t,\tau)\right] \tag{7.21}$$

where g and *e* refer to ground-state and excited-state coherence. For zero vibrational damping

$$C_u(t,\tau) = K_u(t-\tau)\exp\left(i\omega_0 A_u \Delta \int_\tau^t \mathrm{d}s\, \mathrm{e}^{-\gamma|s|} \times \cos(\omega s + \phi_u)\right) \tag{7.22}$$

$C_u(t,\tau)$ is the correlator that includes vibrational coherence, showing modulation at the vibrational frequency ω. In Eq. (7.22) Δ is the dimensionless displacement for the vibrational mode. Note that $\Delta = 0$ implies no modulation, so only Franck–Condon active modes can give rise to coherent phonons. $A_u = \sqrt{\left(Q_{0u}^2 + P_{0u}^2\right)}$ specifies the wavepacket initial position and momentum.

For excited-state wavepacket $P_{0e} = 0$. Accordingly, the coherent vibrational amplitude is $\overline{Q}_g(t) = |A_g| \cos(\omega_0 t + \varphi_g)$ in the ground state, and $\overline{Q}_e(t) = \Delta + |A_e| \cos(\omega_0 t + \varphi_e)$ in the excited state. The initial displacement Q_{0u} and momentum P_{0e} are obtained by integration in the frequency space of the overlap between the absorption spectrum and the excitation pulse spectrum. In the original article by Kumar *et al.* [3], a very convenient series expansion is introduced to obtain an analytical solution that can be worked out in the frequency space, thus avoiding cumbersome numerical evaluation in the time domain. Finally, we provide a quick view to the time evolution expression for the time-dependent correlator for vibrational coherence. This is similar to that in Eq. (7.2), but different in the initial condition, that is now nonstationary. This is a consequence of the effective linear approximation, which describes nonlinear response as linear response from a nonstationary state. For stimulated emission, the expression, as depicted in Figure 7.12 is

$$C_e(t,\tau) = e^{-\gamma(t-\tau)} \left\langle i(\tau) \left| e^{+iH_g(t-\tau)/\hbar} \mu_{eg}^{*} e^{-iH_e(t-\tau)/\hbar} \mu_{eg} \right| i(\tau) \right\rangle \tag{7.23}$$

$i(\tau)$ is the wavepacket initially placed on the excited-state PES after a propagation time τ. Comparing this with Eq. (7.2), one sees that correlation is now worked out between wavepackets after their respective evolution. First, the pump pulse places impulsively a wavepacket onto the excited state, $i(0)$. This wavepacket propagates for a time $\tau, |i(\tau)\rangle = e^{-iH_e(\tau)/\hbar}|i(0)\rangle$. At this point, the probe pulse interacts with the system, bringing part of the wavefunction amplitude to the ground state. The transition is vertical and simply projects the excited-state wavepacket onto the ground-state wavepacket, changing its relative position and phase (momentum) coordinates. Mathematically this simply corresponds to a change of coordinates $x \to (x - \Delta)$. The wavepacket on the ground state evolves for a time $(t - \tau)$ under ground-state Hamiltonian $e^{-iH_g(t-\tau)/\hbar}|i(\tau)\rangle$. The wavepacket left on the excited state evolves in time according to $e^{-iH_e(t-\tau)/\hbar}|i(\tau)\rangle$. Then a second probe field interaction projects the remaining excited-state wavepacket onto the ground-state potential,

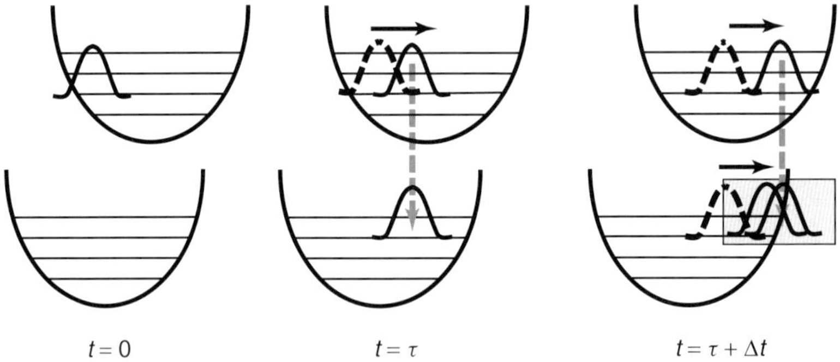

Figure 7.12 Linear response of the nonstationary state in the time domain. The gray rectangular frame indicates when correlation is worked out.

and correlation is worked out. This picture of the process is very useful, and one can work out the transient stimulated emission spectrum by integrating Eq. (7.23) using the analytical propagation for Gaussian pulses, according to Eq. (7.8).

References

1. Simpson, W.T. and Peterson, D.L. (1957) *J. Chem. Phys.*, **26**. 588.
2. Heller, E.J. (1981) *Acc. Chem. Res.*, **14**, 368–375.
3. Kumar, A.T.N., Rosca, F., Widom, A., and Champion, P.M. (2001) *J. Chem. Phys.*, **114**, 701–724.

8
Charge Transfer and Transport

8.1
Adiabatic Electron Transfer: Classical Marcus Theory

Let us first consider a system in equilibrium with a bath, for instance, in a solvent. The system could be a molecule, a pair of molecules, and so on. In the classical approximation we use the total free energy G and a one-dimensional global reaction coordinate x to describe the state of the system. The global coordinate represents all the relevant degrees of freedom of the system, and one can think of it as a linear combination of many contributions, including nuclear coordinates (translational, rotational, and vibrational) and positions and orientations of the solvent molecules. Marcus [1] assumes the linear response approximation, which implies that any change in the system produces a proportional change in the solvent, and thus $G(x) \sim x^2$ (Figure 8.1).

The easiest electron transfer to start with is the self-exchange $A^- + A^0 \rightarrow A^0 + A^-$ (for instance, $Fe^{2+} + Fe^{3+} \rightarrow Fe^{3+} + Fe^{2+}$) between equal donor and acceptor atoms, molecules, or systems. In this example

1) the product is equal to the reactant;
2) no chemical bonds are broken.

Note that assumption (2) is always true in the following discussion.

In 1952, Libby observed that when an electron is transferred from a reacting ion to another, the two new molecules or ions formed are in the "wrong" environment and configuration (i.e., they are out of equilibrium). This means, for instance, that A^0 and A^- have different atomic configurations and different solvent cage polarization. Behind this assumption is the same reasoning that leads us to the Frank–Condon (FC) principle in spectroscopy. Because the transition time is very short, the system does not have time to readjust. Such a vertical transition will not occur in the dark, because of energy conservation. Marcus starts from this observation to develop his theory of the transition state (TS).

For self-exchange we can refer to the plot in Figure 8.1. We distinguish the reactants (R) from the products (P). Even if the two systems are the same, electron transfer leads to a change in the global coordinate, which we represent with a

The Photophysics behind Photovoltaics and Photonics, First Edition. Guglielmo Lanzani.

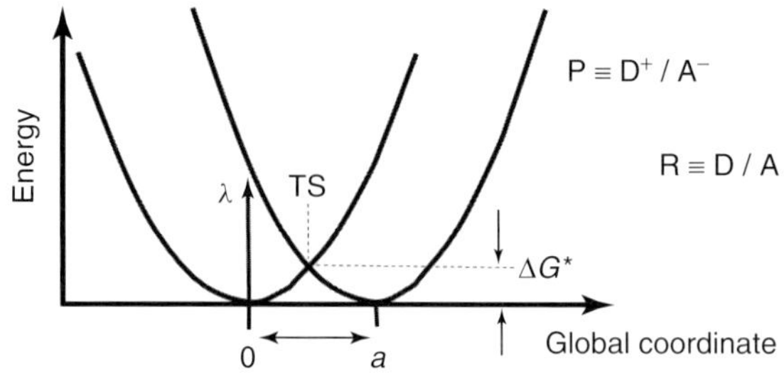

Figure 8.1 Free energy parabola for the reacting and product system, for self-exchange. A vertical transition from "0" cannot occur spontaneously.

displacement a. The two energy surfaces are quadratic in x:

$$y_R = x^2; \quad y_P = (x - a)^2 \tag{8.1}$$

They intersect at the TS where they are degenerate, $y_R = y_P$. A very simple mathematical analysis leads to the result for the barrier height, after working out the TS abscissa, $\Delta G^* = \left(\frac{a}{2}\right)^2 = \frac{a^2}{4} = \frac{\lambda}{4}$. The quantity λ is the "reorganization energy." This is the energy required to distort the configuration of the reactant into the configuration of the product without electron transfer. To get to an expression for the rate of electron transfer, Marcus uses the Arrhenius relationship:

$$K_{ET} = Ae^{-\frac{\Delta G^*}{kT}} = Ae^{-\frac{\lambda}{4kT}} \tag{8.2}$$

A is a pre-factor depending on the nature of the electron transfer reaction, for instance, an encounter probability for bimolecular reactions. It usually expresses an attempt frequency that provides the proper scaling for the rate. A also has a weak temperature dependence and is usually determined experimentally.

The mechanism behind electron transfer is as follows: fluctuations in x lead the system to the TS where a quasi equilibrium between reactant and product is postulated. For this reason, using equilibrium statistical mechanisms, the rate expression can be written as a thermal activated hop. At TS, electron transfer can occur, preserving energy conservation and without system distortion. The transfer is followed by relaxation to the equilibrium configuration of the product.

Now we consider an electron transfer reaction between a donor (D) and an acceptor (A) that are not necessarily the same. Let R $=$ D/A, and P $= D^+/A^-$. We assume equal curvature for the total free energy on x for both D and A. Because the systems are in general different, there can be a further gain in energy on electron transfer. For instance, this occurs in photovoltaics when a driving force due to different electronic potential occurs at the donor–acceptor (DA) interface. We call this energy $\Delta G_0(<0)$ (Figure 8.2). Again, straightforward mathematics leads to the following result for the coordinates of the TS state.

$$x_{TS} = \frac{a^2 + b}{2a}; \; y_{TS} = \frac{\left(a^2 + b\right)^2}{2a} = \frac{\left(\lambda + \Delta G^0\right)^2}{4\lambda} = \Delta G^* \tag{8.3}$$

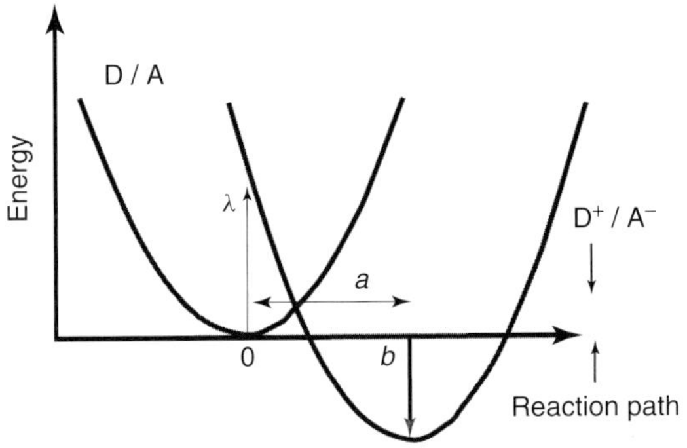

Figure 8.2 Vertically displaced free energy surfaces. The shift $b = -\Delta G_0$.

As before, we can write the rate of exchange as $K_{ET} = Ae^{-\frac{\Delta G^*}{kT}}$. This expression completely characterizes the electron transfer reaction in terms of three quantities: A, ΔG_0, and λ. The barrier ΔG^* is

$$\Delta G^* = \frac{\lambda}{4}\left[1 + \frac{\Delta G^0}{\lambda}\right]^2 \tag{8.4}$$

The reorganization energy plays a crucial role in defining the electron transfer rate. Above it has been defined in very general terms; however, it can be further specified by "internal" (λ_i), and "external" (λ_o) components. The external contribution is worked out within the two hard spheres model and considering the electrostatic energy balance. This corresponds to the thermodynamical average of interaction of the solvent and the bath with the system

$$\lambda_o = \left[U_e^\infty(1) + U_e^\infty(2) - U_{12}^\infty\right] - \left[U_e^0(1) + U_e^0(2) - U_{12}^0\right] \tag{8.5}$$

$$U_e^k(j) = \frac{\Delta e^2}{8\pi\varepsilon_k a_j} \tag{8.6a}$$

$$U_{12}^k = \frac{\Delta e^2}{4\pi\varepsilon_k R} \tag{8.6b}$$

with $k = \infty$ or 0 and $j = 1, 2$; a_j are ionic radii, and R is the separation between the centers of the spheres. The first two terms in each square bracket are the cost in energy to realize the ion/neutral molecule configuration; the last term is the electrostatic interaction between the ions. The first square bracket assumes the instantaneous response of the solvent bath, screened by the medium *polarizability*. The second square bracket represents the cost in energy and the interaction energy at equilibrium, when the solvent *polarity* comes into play.

Because the electron transfer is assumed to take place on a very fast time scale, the bath can respond only with the "instantaneous" electronic polarizability, related to n^2 at optical frequencies. At a later time, when equilibrium is reached, it is the polarity contribution that matters, weighted by the static dielectric function.

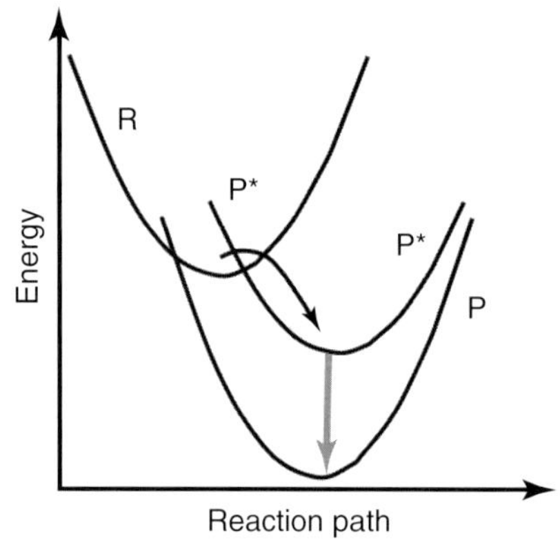

Figure 8.3 The Marcus inverted region. Probability for R to P* is higher than for R to P, so electron transfer generates the excited state P*, which decays radiatively to P.

Remember that there can be a very large difference between n^2 at optical frequency and $\varepsilon(0)$. For instance, in water, $n^2 = \varepsilon_\infty \approx 2.2, \varepsilon_0 \approx 80$.

As an example, we estimate the outer (solvational) reorganization energy for the self-exchange electron transfer reaction Fe^{2+}/Fe^{3+} in water. The result is $\lambda_o = \frac{\Delta e^2}{4\pi\varepsilon_0}\left[\frac{1}{2r_D} + \frac{1}{2r_A} - \frac{1}{r_{DA}}\right]\left(\frac{1}{n^2} - \frac{1}{\varepsilon_r}\right) \cong 1,2$ eV by using $n \approx 1{,}3, \varepsilon_r \approx 80, r_{DA} = 2r_D = 2r_A \approx 0.7$ nm.

Note finally that for nonpolar solvent $\varepsilon_{opt} = n^2 \approx \varepsilon_r$ and $\lambda_o \to 0$.

The internal contribution to the reorganization energy is specified by elastic energy of the chemical bond rearrangement as $\lambda_i = \frac{1}{2}\sum_j K_j\left(Q_j^R - Q_j^P\right)^2$, where $Q_j^{R,P}$ are equilibrium values for the jth mode in R or P and K_j is a reduced force constant $K_j = \frac{2K_j^R K_j^P}{K_j^R + K_j^P}$ where $K_j^{R,P}$ are force constants for mode j in R or P.

An interesting prediction of the Marcus theory is the existence of an inverted region, when the downward energy gap is very large and the rate gets smaller. This apparently counterintuitive behavior is well understood by looking at the two potential surface crossing plots. For very large downshifts in energy of the product curve, the barrier for hopping goes through zero and then grows again. The inverted region has been experimentally demonstrated. This was a big success of Marcus theory and an important validation. In addition, the Marcus inverted region accounts for well-known daily life phenomena such as the safety light rod. When the rod is bent a glass ampoule is broken, the liquids from the ampoule and the rod are mixed and "cold light" is produced. The phenomenon is an example of chemiluminescence, a complicated process involving electron transfer steps. Owing to the inverted region, the total reaction results in an excited state from which light is emitted. Figure 8.3 well explains this. The electron transfer to the excited state is more probable than that to the ground state. Safety lights of this kind, which are nonflammable and weatherproof, are used by seamen and divers in emergency.

8.2 Quantum Mechanical Expressions for Electron Transfer: Nonadiabatic Multiphonon Regime

We distinguish two regimes of electron transfer: adiabatic and nonadiabatic. In the classical picture, we see the two potential surfaces crossing at TS. In quantum mechanics, this degeneracy is removed by a splitting, Δ, which is twice the interaction energy, as seen in Figure 8.4. If splitting is large with respect to the average phonon energy, we are in the adiabatic regime. When the splitting is small, we are in the non-diabatic regime. In the adiabatic case, electron transfer occurs through the TS along the reaction coordinate that involves a concerted motion of the nuclei. During adiabatic transfer, the electron remains in the same Born–Oppenheimer (adiabatic) state, changing continuously the localization along the reaction coordinate. The adiabatic rate is typical for classical phenomena and forms the basis of the TS theories in which the reaction rate is determined by the probability of finding and crossing the TS. This is the typical Marcus regime.

The nonadiabatic effect reduces the probability of electron transfer that happens at TS but opens up a new channel involving direct transitions from D to A that can occur at any nuclear configuration. Such transitions are nonradiative transitions between vibronic states of D and A, which become important when the D–A coupling is weak. The thermally averaged transition rate can be adequately described by the first-order perturbation theory, using the Fermi golden rule, where the rate of transfer is proportional to the density of acceptor states. The nonadiabatic transfer is a quantum effect that better accounts for the temperature dependence, especially at low T and shows an exponential decay on D–A separation.

According to the Fermi golden rule the electron transfer rate is given by

$$k_{\mathrm{ET}} = \frac{2\pi}{\hbar} V^2 FC \tag{8.7}$$

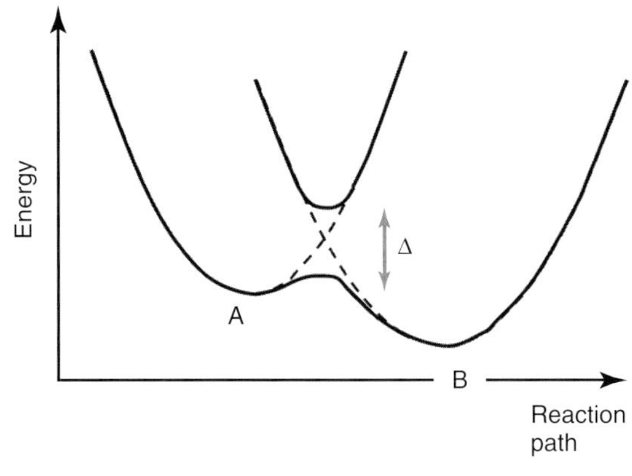

Figure 8.4 Dashed lines are crossing of potential surfaces according to the classic picture. Solid lines are adiabatic potential surfaces showing quantum splitting Δ.

Here FC is the Frank–Condon overlap factor as defined in previous chapters; V is the coupling strength, usually described by an exponential decay with distance according to $V^2 = J_0^2 \exp[-\beta(r - r_0)]$, where J_o is an appropriate matrix element;, β is the scale factor; and r_0 is the minimum distance dictated by the steric hindrance of the interacting units. Details of the calculation can be found in the original work of Joshua Jortner [2]. In his theory, there are two classes of vibrations: those with very low frequencies, which are due to the solvent, and those with high frequencies, proper of the interacting systems (intramolecular). In general, a vibration can be treated as classical if $\hbar\omega_{\text{vib}} \ll kT$; it should be treated as quantum if $\hbar\omega_{\text{vib}} \gg kT$. We can distinguish three temperature regimes that give a different description of the electron transfer rate. In the intermediate temperature range, the most common, the solvent vibrational modes can be treated classically, while the molecular vibrations are quantum, that is, $\hbar\omega_s \ll kT$ (where ω_s is an average frequency) and $\hbar\omega > kT$. The solvent overlap integrals follow a Gaussian distribution, and the rate of ET is given by

$$k_{\text{ET}} = \frac{2\pi}{\hbar} V^2 \left(\frac{1}{2\pi\lambda_0 kT}\right)^{1/2} e^{-S} \sum_{m=0}^{\infty} \exp\left[-\frac{(\lambda_0 + m\hbar\omega + \Delta G_0)^2}{4\lambda_0 kT}\right] \frac{S^m}{m!} \tag{8.8}$$

where the high-frequency phonon modes are specified by a mean frequency ω and by an effective electron–phonon coupling strength $S = \sum_k \Delta_k^2/2$ where $\{\Delta_k\}$ are dimensionless configurational displacements. Equation (8.8) resembles the classical one. The λ_0 energy term in the original work of Jortner accounts only for the solvent vibrational reorganization energy, $S_S\hbar\omega_S$, while the electrostatic energy contribution is not considered. Here, however, there is explicit temperature dependence, and the nonadiabatic paths through tunneling are weighted by a probability proportional to the square vibrational overlap integrals for intramolecular (high-frequency) modes in harmonic approximation. This formalism is similar to that adopted for nonradiative transitions and also for vibronic absorption. When both vibrational systems can be treated classically (high temperature limit), the expression for nonadiabatic multiphonon electron transfer becomes the conventional activated rate equation:

$$k_{\text{ET}} = \frac{2\pi}{\hbar} V^2 \left(\frac{1}{2\pi\lambda kT}\right)^{1/2} \exp\left[-\frac{(\lambda + \Delta G_0)^2}{4\lambda kT}\right] \tag{8.9}$$

where the FC factor is described by the exponential term. Again, according to Jortner the reorganization energy is $\lambda = S_S\hbar\omega_S + S\hbar\omega$.

For very low temperature, $\hbar\omega \gg \hbar\omega_S \gg kT$, both vibrations should be treated quantum mechanically and the expression is

$$k_{\text{ET}} = \frac{2\pi}{\hbar^2\omega_s} V^2 e^{-(S_S+S)} \sum_{m=0}^{\infty} \frac{S_s^{p(m)} S^m}{p(m)!m!} \tag{8.10}$$

where the low-frequency phonon modes of the exterior medium are specified by a mean frequency ω_s, reduced displacements $\{\Delta_{Sk}\}$, and an effective coupling $S_S = \sum_k \Delta_{Sk}^2/2$. This physical situation corresponds to temperature-independent tunneling between the zero point of the nuclear configuration of the initial state to the vibronic states of the final nuclear surface, which are nearly degenerate with it. $p(m)$ is the average number of solvent phonon modes involved in the transition (integer of $\frac{-\Delta G_0 - m\hbar\omega}{\hbar\omega_S}$).

In polar liquids, $\hbar\omega_S \approx 1$–$10\ \mathrm{cm}^{-1}$, and in a molecular solids $\hbar\omega_S \approx$ 10–100 cm^{-1}(describing both intermolecular phonons or solid matrix phonons). Intramolecular frequencies can range between 300 and 3000 cm^{-1}.

8.3 The Donor–Acceptor Interface

Electron transfer implies extraction of one electron from the donor and its assignment to the acceptor. The energy required for removing one electron from the neutral molecule is called ionization energy (IE), and the energy gained by adding one electron to the neutral molecule is called electron affinity (EA). Both are differences between an initial state energy and a final state energy. For ionization in vacuum, the energy of the initial state is that of the neutral molecule, E^0, and the final state energy is the sum of the cation energy, E^+, and the energy of a free electron at rest, infinitely far away from the charged molecule, E_{vac}. Usually $E_{vac} = 0$ and $IE = E^+ - E^0$. For EA, the final-state energy is that of the anion, E^-, and the energy of the initial state is the sum of E_0 and E_{vac}; because EA is a positive quantity, $EA = E^0 - E^-$. Differences $E^+ - E^0$ and $E^0 - E^-$ are often approximated by the energies of the highest occupied molecular orbital (HOMO) and the lowest unoccupied molecular orbital (LUMO), respectively. So, disregarding correlation, a single isolated molecule has one and only one well-defined $IE \approx -E_{\mathrm{HOMO}}$ and $EA \approx -E_{\mathrm{LUMO}}$.

Continuing our simple reasoning, the cost of electron transfer in a simple coordinate-independent model is $\Delta G_0 = IE - EA + E_{\mathrm{CT}}$, where $E_{\mathrm{CT}} < 0$ is the Coulomb binding energy of the displaced charges (the electron and the hole left behind). In a crystal, $\Delta G_0 = IE - EA + E_{\mathrm{CT}} + \Delta P$, where ΔP accounts for the lattice distortion and electrostatic polarization around the two ions. According to Marcus theory, E_{CT} and ΔP are included in the reorganization energy, λ. Using $IE \sim 5$ eV, $EA \sim 3$ eV, $\varepsilon \sim 3$, and $\mathrm{r} \sim 0.5$ nm (for dielectric constant and charge separation), neglecting ΔP, $-\Delta G_0 \sim 1$ eV.

In his Nobel Prize lecture, Marcus comments on the ever-expanding application of the electron transfer model to many research areas. In particular, he mentions biology and photosynthesis, the process that is now inspiring a whole community of researchers looking to improved photovoltaic conversion devices. In photosynthesis, energy harvested from solar radiation is transferred from antenna chlorophylls to the special pair BChl2. The latter then transfers an electron to a pheophytin BPh within about 3 ps and from it to a quinine Q_A in about 200 ps and thence to another quinine Q_B. Nature exploits systems to accomplish functions, and the reaction chain is extremely rich and complex. On the basis of his theory, Marcus suggests a guide to have highest efficiency in three steps. (i) To avoid wasting excitation energy of the $BChl_2^*$ it is necessary to have a small $-\Delta G_0$ in the first electron transfer to BPh. The estimated value is 0.25 eV, of an overall excitation of 1.38 eV. (ii) For high efficiency against competing channels that waste energy such as fluorescence and internal conversion of $BChl_2^*$, it is necessary that ΔG^* be small in that electron transfer and the reorganization energy be small. Because the size of the reactants is large and the protein environment is nonpolar, this is well accomplished in nature. (iii) Unwanted back electron transfer reaction BPh^- to $BChl_2^+$ should be slow to allow the second step to take place (transfer top Q_A). This seems difficult because electron–hole recombination is largely exothermic (about 1.1 eV). It thus appears that the small reorganization energy and the resulting inverted region plays a role in providing the essential conditions for an effective reaction. Amazingly, experiments show that natural systems are indeed finely tuned according to these rules, validating Marcus theory.

8.4
Charge Photogeneration in Excitonic Semiconductors

The quantum yield of charge photogeneration is $\eta = \frac{n^{\circ}(e-h)}{n^{\circ}\text{photons}}$, where (e–h) are free carriers and the denominator accounts for absorbed photons. In bandlike semiconductor $\eta = \eta(h\nu, T) \xrightarrow[k_B T \gg E_b^X]{} 1$ because the exciton is ionized at high-enough T. For high T or photon energy exceeding the valence-conduction band gap we can write $\eta \approx 1$. The process of e–h generation has been discussed in Section 4.1.

In amorphous semiconductors, molecular solids, or low-dimensional solids, the zero field charge photogeneration yields for homogeneous samples is 10^{-3}–10^{-5}. External actions, such as electric field, temperature, disorder, and interfaces enhance this figure, sometime to unity.

In such materials photogeneration is a multistep process. A possible description is

1) Photon absorption and generation of a neutral singlet state S_n;
2) Auto-ionization (AI) with generation of hot carrier pair;
3) Thermalization and generation of an intermediate, bound charge pair;
4) Dissociation of the intermediate pair into free carriers.

$$S_0 \xrightarrow{h\nu} S_n \xrightarrow{\phi_0} CT \xrightarrow{\Omega} \begin{cases} (+) \\ (-) \end{cases} \quad \eta = \phi_0 \Omega$$

Figure 8.5 shows the multistep process. We discussed in detail process (1) in previous chapters. Here we just note that it will make a big difference in terms of later developments if the initial neutral state, represented as a molecular singlet state in Figure 8.5, is indeed a Frenkel exciton, a Wannier exciton, or a CT state. Engineering of the photovoltaic material should carefully take into account the nature of the initial state.

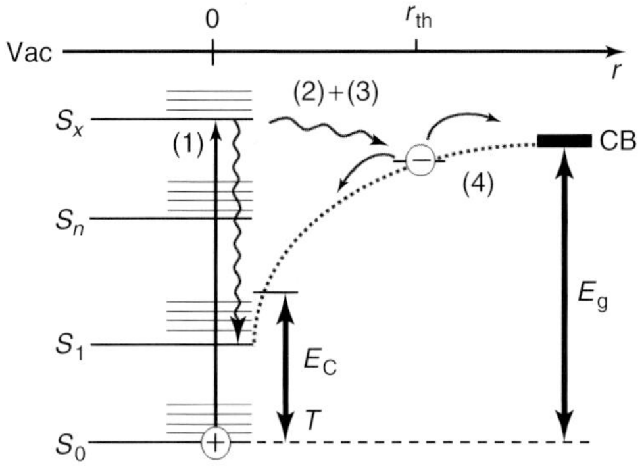

Figure 8.5 The multistep charge photogeneration process. (1) Absorption, (2) autoionization, (3) thermalization, and (4) dissociation.

AI is a subtle phenomenon, regarding a neutral state that breaks into a pair of charges, on a very short time scale and usually without external perturbations. The name was proposed in the 1940s to explain the small probability to generate charge pairs from tightly bound neutral states in molecular crystals. It is sometimes described as a scattering event from a discrete level into a continuum (the energy band of the charge pair). Theory describes it by first considering the coherent superposition of neutral and charged states and then assuming a different phonon-induced dephasing for the two types of states. Initially the electronic population oscillates between the two types of states, coupled by Coulomb interaction. The different coupling to the phonon bath induces different rates (probability) of decay, thus favoring stabilization of one of the two. Alternatively, a transition rate between neutral and charged states can be introduced, because of Coulomb interaction acting as a perturbation. The wavefunction character of the initial state and its overlap to the CT wavefunction plays a role in determining the AI rate. Whatever the microscopic mechanism, AI is the ultrafast appearance of a pair of charge carriers after absorption into a neutral state (Figure 8.5). The process can be characterized by a rate k_{AI} and efficiency according to

$$\Phi_0 = \frac{k_{\mathrm{AI}}}{k_{\mathrm{AI}} + k_n} \tag{8.11}$$

where k_n contains all other deactivation paths for the state S_n. For amorphous inorganic semiconductors ϕ_0 is about 1, while in homogeneous molecular semiconductors it is much smaller. AI can be intermolecular or intramolecular. For small molecules it is of the first kind, but for conjugated segments in polymer chains it can be of the second kind. Essentially, it is a matter of "space" to accommodate the charge-separated pair. A clear picture for AI comes from the DA pair (Figure 8.6). In this case, the initial neutral state of the donor experiences a driving potential because of the interface with the acceptor and breaks into a charged pair by electron or hole transfer. In this situation, the efficiency of AI can reach unity also in molecular solids.

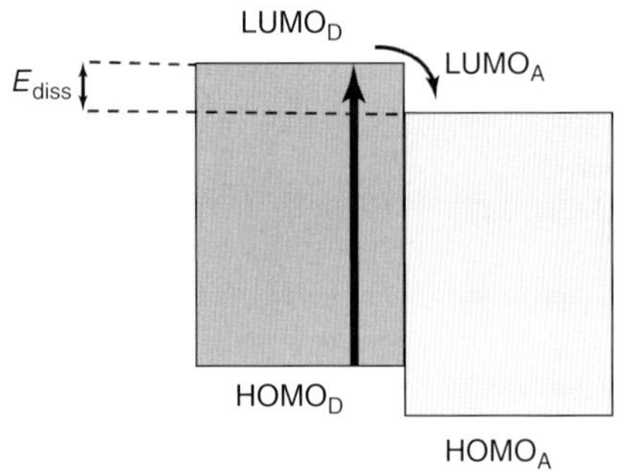

Figure 8.6 Example of autoionization at donor–acceptor interface.

Thermalization concerns the dissipation of some excess energy toward a relaxed e–h pair. AI takes place above a threshold. The excess energy goes into phonons and leads to a local increase in temperature, once the phonon population equilibrates. The excess energy $\Delta E = h\nu - E_{CT}$ depends on the photon energy, while E_{CT} is the minimum energy for creating a CT state. Note that in excitonic or disordered semiconductors this energy is usually larger than the optical gap (the minimum energy for photon absorption), while in bandlike semiconductors the optical gap E_g coincides with the electrical gap, E_{CT}. The thermalization time, τ_{th}, can be estimated within the semiconductor model as the time needed to transfer the excess energy to the phonon bath. Assuming that the maximum rate of phonon emission is the phonon frequency, ν_p, and considering that each phonon takes away an energy $h\nu_p$, we get $\tau_{th} \geq \frac{\Delta E}{h\nu_p^2}$. Assuming an excess energy of 1 eV and a phonon frequency of $2 \times 10^{13}\ s^{-1}$, $\tau_{th} \approx 600$ fs. During this time, the electron ejected from the initial state can wander around and end up at an average distance equal to the mean free path under diffusion, $r_{th} = \sqrt{D\tau_{th}}$. Using $D = 1\ cm^2/s$, this gives $r_{th} \approx 10$ nm. Merging together those two equations, one gets an estimate of the thermalization distance (the distance between positive and negative charges after thermalization) of $r_{th} \cong \left(\frac{D \cdot \Delta E}{h\nu_p^2}\right)^{1/2}$. This suggests that (i) the higher the excess energy the farther away the charges thermalize and the less bound will be the CT state and (ii) better mobility implies a higher diffusion constant and thus a larger distance for the CT pair. Once the CT is cooled off, the activation energy to be overcome for freeing the charges is $E_A = \frac{e^2}{4\pi\varepsilon r_{th}}$. This suggests introducing a Coulomb capture radius, which is the balance between the Coulomb and the thermal energies, according to $r_C = \frac{e^2}{4\pi\varepsilon KT}$. A pair of charges separated by a distance $r \gg r_C$ will be free simply by thermal diffusion, while the opposite situation, $r \ll r_C$ represents a bound CT state.

Dissociation can be described by the Onsager model. For $r < r_C$ Onsager assumes that the charge carrier undergoes a random walk in the presence of Coulomb attraction to the opposite charge and external electric field E. The Coulomb potential is $U(r) = -\frac{e}{4\pi\varepsilon r} - eEr\cos\theta$, where the dielectric constant is the only way to account for the solid. The Onsager model was initially developed for describing ions in a liquid, and it is indeed quite a strong approximation for a semiconducting solid, yet it provides a good framework for understanding charge separation phenomena. The mathematics of the model is quite complex. The rate equation for the density of diffusing particles is

$$\frac{\partial n(\bar{r},t)}{\partial t} = \frac{kT}{e}\mu\nabla\cdot\left(\mathrm{e}^{-U/KT}\nabla\left(n\mathrm{e}^{U/KT}\right)\right) \tag{8.12}$$

and $r = 0$ is a sink where recombination takes place. The initial condition is $n(r,0) = g(r_{\mathrm{th}})$ where "g" is a distribution function of the initial thermalization distance. The model allows working out the escape probability Ω. The *escape probability* is defined as the probability of forming free carriers from the initial thermalized CT pair, $\Omega = \frac{k_{\mathrm{d}}}{k_{\mathrm{r}}+k_{\mathrm{d}}} = k_{\mathrm{d}}\tau_{\mathrm{CT}}$ where k_{d} is the rate of dissociation, k_{r} is the rate of recombination and τ_{CT} the lifetime of the CT. Ω depends on the electric field strength, on the initial thermalization distance, and on the dimension of the space wherein random walk is taking place. The solution of the Onsager model for an isotropic distribution of initial pairs, peaked at r_{th} and in 3D space, is

$$\Omega^{3\mathrm{D}} = \mathrm{e}^{-r_{\mathrm{c}}/r_{\mathrm{th}}}\left[1 + \frac{e}{KT}\frac{r_c E}{2!} + \cdots\right. \tag{8.13}$$

For weak field this can be approximated by $\Omega^{3\mathrm{D}} = A(T)\mathrm{e}^{-r_{\mathrm{c}}/r_{\mathrm{th}}} \approx \Omega_0$, where $A(T)$ is a weakly temperature-dependent function. In this case, the escape probability is independent from the electric field. Note that the exponential factor in Eq. (8.13) has the simple Boltzman expression $\mathrm{e}^{-E_a/kT}$. This represents the thermal population fraction of dissociated pairs. According to Braun, the Onsager equation should be corrected to take into account the dynamic balance between CT dissociation and CT regeneration, $\mathrm{CT} \underset{\gamma}{\overset{k_{\mathrm{d}}}{\rightleftarrows}} \mathrm{P}^+ + \mathrm{P}^-$. Here, γ is the bimolecular recombination constant of the free carriers. At equilibrium, the rates of dissociation and bimolecular recombination are equal, $k_{\mathrm{d}} = \gamma\Omega N_{\mathrm{CT}}$, where the dissociated pair population density is ΩN_{CT}, assuming Onsager is a valid model. The bimolecular recombination constant can be expressed according to Langevin model, $\gamma = \frac{\mu e}{\varepsilon}$, to get $k_{\mathrm{d}}(F) = \frac{\mu e}{\varepsilon} N_{\mathrm{CT}}\Omega(F)$. The crucial difference is that in the Braun model, pair recombination leads to CT regeneration and subsequent redissociation, while in the Onsager model the recombined pair disappears from the system. Braun model with its modified dissociation rate is consistent with thermalization distances sizably smaller than those needed when fitting the original Onsager model. The explicit dependence on mobility also suggests that better transport might help dissociation of the initial CT pair.

For increasing field, the escape probability becomes linear with the field, $\Omega^{3\mathrm{D}} \propto E$, until saturation at unity (Figure 8.7). The critical field is $E \approx \frac{e}{4\pi\varepsilon r_{\mathrm{th}}^2}$.

With respect to r_{th}, we can distinguish two regimes:

$$r_{\mathrm{th}} < r_{\mathrm{C}} \quad \Omega^{3\mathrm{D}} = \Omega^{3\mathrm{D}}(F) \tag{8.14a}$$

$$r_{\mathrm{th}} > r_{\mathrm{C}} \quad \Omega^{3\mathrm{D}} \sim 1 \tag{8.14b}$$

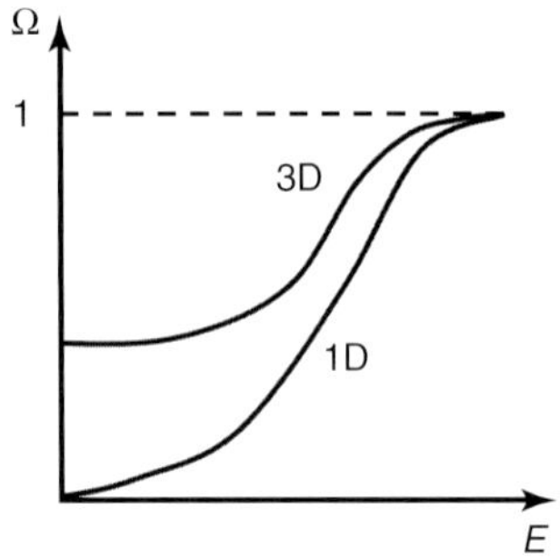

Figure 8.7 The escape probability according to 1D and 3D Onsager model versus electric field amplitude.

This is an intuitive result, telling us that large photon energy provides excess energy, enough to overcome the Coulomb attraction, and thus the escape probability tends to unity.

Dimensionality has an important role. Remember that a random walker in 1D always goes back to the origin, with time-dependent survival probability $\propto t^{-1/2}$. In 1D there is no thermal escape, and the electric field becomes essential for dissociation. The escape probability can be cast into a simple statistical form:

$$\Omega^{1D} = \frac{\int_0^{r_{th}} \exp\left(\frac{eU}{k_B T}\right) dx}{\int_0^{\infty} \exp\left(\frac{eU}{k_B T}\right) dx} \tag{8.15}$$

And for $r_{th} < r_C$, $E < \frac{k_B T}{e r_C}$ the solution is $\Omega^{1D} = \frac{e r_{th}^2 E}{r_C KT} e^{-r/r_{th}} \propto E$. Note that for zero field the escape probability is zero. Figure 8.7 summarizes the considerations.

Poole–Frenkel is another model for the escape probability, considering thermal hopping above a Coulomb barrier lowered by the external field. The Poole–Frenkel model is suitable for weakly bound pairs.

According to Figure 8.8, the barrier height is $E_b = E_{CT} - \Delta E_0 = E_{CT} - \beta F^{1/2}$, where $\Delta E_0 = \sqrt{\frac{e^3 F}{\pi \varepsilon}}$, corresponding to a distance $r_m = \sqrt{\frac{e}{\varepsilon E}}$. The escape rate according to this model is

$$k_{PF} = \nu \exp\left(-\frac{E_{CT}}{KT}\right) \exp\left(\frac{\beta F^{1/2}}{KT}\right) \tag{8.16}$$

where ν is an attempt frequency, typically associated with a promoting phonon. k_{PF} is the product of a field-independent dissociation rate and a field-dependent hopping rate. The value of E_{CT} represents the binding energy of the CT state. For well-separated pairs, it can be approximated by the Coulomb binding energy, coinciding with the activation energy above, $E_A = \frac{e^2}{4\pi\varepsilon r_{th}}$. The preexponential factor can be experimentally estimated, while it is difficult to be predicted by theory. Experiments suggest a value of 10^{14}–10^{15} s^{-1}.

The escape probability is $\Omega_{PF} = \frac{k_{PF}}{k_{PF} + k_{GPR}}$, where k_{GPR} is the geminate pair recombination rate.

Recent experiments also show that many conjugated polymers follow the rule that larger photon energy leads to higher charge photogeneration yield, pointing to hot

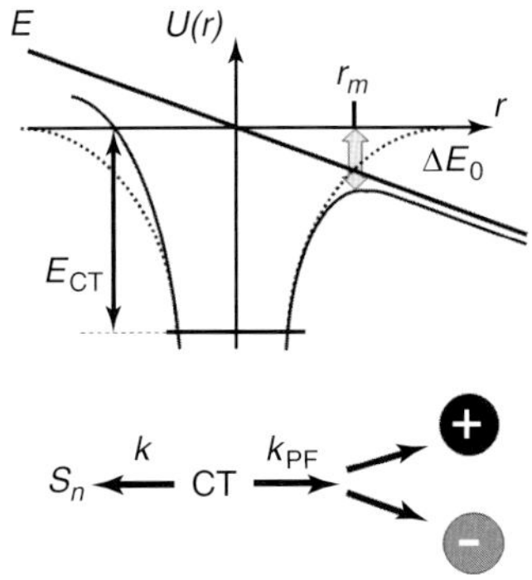

Figure 8.8 Poole–Frenkel model: field-assisted ionization through a Coulomb barrier.

state dissociation. However, this is not a general behavior. "Cold" singlet exciton dissociation is also observed, for instance, in photoconductors, such as phthalocyanine or others developed for the photocopy process, which follows the Vavilov–Kasha rule. Here internal conversion is fast, 10^{-13}–12^{-12} s, and following photoexcitation to any excited state leads to filling up of the lowest excited state, S_1, which is long lived, 10^{-9}–10^{-8} s. Observation of excitation-wavelength-independent photoconductivity suggests that dissociation may still occur as field-assisted or thermal-assisted tunneling from S_1. In this case, the threshold for charge generation follows closely that of optical absorption. The rate of generation of the initial CT state can be obtained by integration over the solid angle of the tunneling probability,

$$K_F = \frac{1}{2}K_0 \int K_F^{\theta} d\Omega_{\theta} = \frac{1}{2}K_0 \int_{-1}^{1} e^{\beta E \cos\theta} d(\cos\theta) = K_0 \frac{senh(\beta E)}{\beta E} \tag{8.17}$$

where both K_0 and β are parameters to be adjusted fitting experimental data. Once the CT state is formed, dissociation may be described by Poole–Frenkel model. The process is depicted in Figure 8.9, compared to the Onsager model. AI is usually considered nonfield dependent, while the further escape, be Onsager or Poole–Frenkel, is. On the contrary, the last mechanism we described is field dependent from the very beginning. In amorphous semiconductors owing to disorder, equal molecules might end up in pairs with level alignment favorable for electron transfer (type two interface). The electric field may assist such configuration and the resulting charge transfer, but it is not strictly required when there are favorable energetic conditions (Figure 8.10). Further separation is usually field or thermal assisted, as it is in a bulk heterojunction cell or a DSSC cell. The process of disorder-induced charge separation is a *hot dissociation* regarding population distribution in the disorder density of states (DOS), because it occurs during migration. Once the system is fully relaxed to the bottom of the DOS, the probability to have a D-A-like level alignment is negligibly small. Another interesting observation is that ultrafast internal conversion from S_n to S_1 may lead to a buildup of local vibrational energy, which could behave as a heat sink for charge dissociation. This model was proposed by Archipov for conjugated polymers. In an article that appeared almost at the same time as that of Popovic about cold dissociation, Braun suggested his modification of the Onsager model

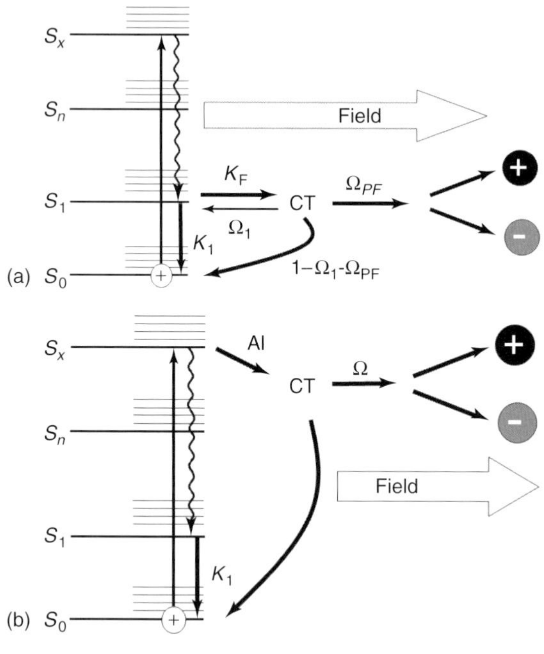

Figure 8.9 (a) The theory of cold dissociation by Z. D. Popovic-J [3] and Noolandi and K. M. Hong [4]. The large arrow indicates processes affected by the field. (b) The theory of hot dissociation of Onsager. The large arrow indicates processes affected by the field.

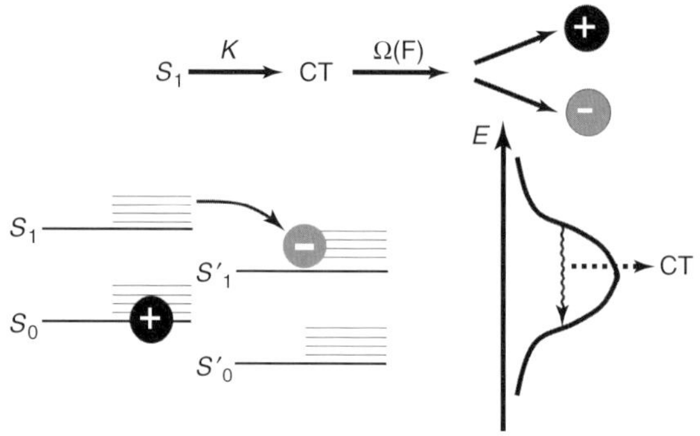

Figure 8.10 Model of D–A pair, eventually induced by disorder to explain charge separation in homomaterials.

to keep into account regeneration of S_1 due to charge carrier recombination. The latter is neglected in the Popovic model. So there are essentially few general ideas that provide a framework for understanding multistep charge generation, while a detailed microscopic theory that can predict quantitative rates does not exist. Experiments are thus crucial. Table 6.11.1 sums up the models we described.

Inorganic crystals	Amorphous semiconductors	Molecular photoconductors
$\eta \approx 1$	$\eta \ll 1$	$\eta < 1$
F NO	F Yes	F Yes
λ NO($<\lambda_{gap}$)	λ Yes	λ NO
PC $\propto \alpha(\omega)$	PC $\neq \alpha(\omega)$	PC $\propto \alpha(\omega)$
VB $\rightarrow$ CB	$S_0 \xrightarrow{h\nu} S_n \xrightarrow{\phi_0} \text{CT} \xrightarrow{\Omega} \begin{cases}(+)\\(-)\end{cases}$	$S_0 \xrightarrow{h\nu} S_1 \xrightarrow{\phi_0} \text{CT} \xrightarrow{\Omega} \begin{cases}(+)\\(-)\end{cases}$

It is instructive to conclude by listing the main charge photogeneration mechanisms that have been proposed for conjugated polymers.

1) Valence band to conduction band transition. This implies the direct instantaneous generation of free carriers. According to this model, however, the electron and hole have strong coupling to the one-dimensional lattice, decaying within a phonon period into self-trapped states such as solitons and polarons, according to the space topology.
2) Direct excitation of a precursor intrachain state, which subsequently dissociates into an interchain geminate pair. The initial state has a peculiar charge distribution that leads to a higher probability of AI.
3) Dissociation of hot photoexcitations. Before vibrational cooling, the initial exciton has enough energy to split into an interchain CT state.
4) Disorder-induced dissociation during interchain migration. The electron transfer is interchain.
5) On chain dissociation, assisted by local heating, followed by interchain hopping. This mechanism finds support from ultrafast pump probe measurements in isolated chains (See Box 8.1).

8.5
Synopsis on Transport

This paragraph deals with charge transport in semiconductors. It is a general and simple introduction. More detailed theories can be found in specialized monographs or books as reported in the bibliography. The main purpose is to provide a basic understanding of the phenomena and the experimental techniques, because photophysics often intercepts transport physics and because in photovoltaic devices transport plays a crucial role, as it does in many other optoelectronic devices.

First we distinguish drift from diffusion.

Drift is the motion of charge carriers under the action of an electric field. It matters, for instance, in polymer bulk heterojunction cells, where an internal field is needed to strip the charge carriers out of the device. The equation for drift

Box 8.1: Charge Generation in Polymer Networks

Lower lying photoexcitations in long conjugated polymer chains are neutral states that can be associated with 1D Wannier excitons. With respect to interchain interactions, much weaker than intrachain interactions, the state is described as a Frenkel exciton for order arrays or as a localized molecular state in amorphous networks. This opens a conceptual gap between the intrachain space and the interchain space. Coulomb attraction between opposite charges cannot be screened in spite of the high electron density along the chain, because of the 3D nature of the electrodynamic interaction (i.e., the interaction occurs through free space around the chain). Free charges can be generated only with large excess energy with respect to the optical gap, or, in other terms, the exciton binding energy is large. The figure shown below sums up a possible model for charge generation, including intrachain and interchain dynamics. It is assumed that high above, gap excitation generates free or weakly bound charges, which can quickly relax into bound exciton states or jump to the nearby chain. Electron transfer leads to an interchain bond pair, or polaron pair, whose further separation may be described by the Onsager framework. Ultrafast pump probe data suggest the lifetime of the initial intrachain charge-separated state to be in the order of a few hundred femtoseconds.

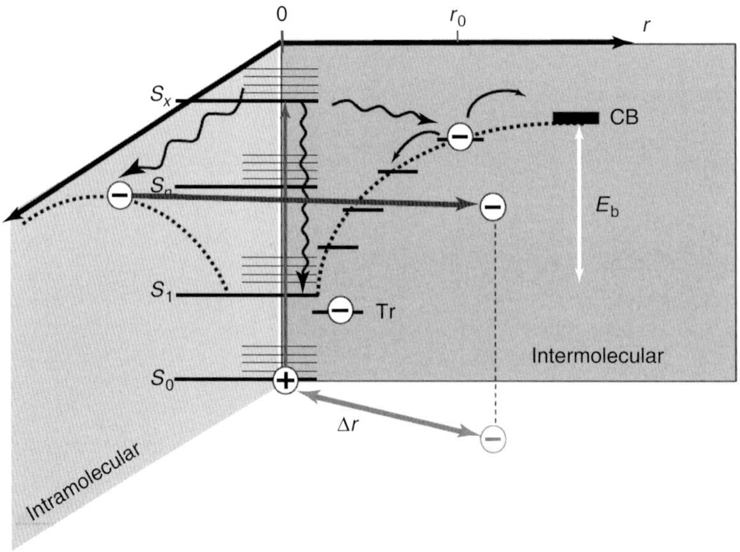

density current is

$$\bar{J} = qn\bar{v}_D = qn\bar{\bar{\mu}} \cdot \bar{E} \tag{8.18}$$

Here, q is the electrical charge of the carrier (C), n is the density of carrier (cm^{-3}), and the drift velocity $\bar{v}_D = \bar{\bar{\mu}} \cdot \bar{E}$, where the tensor quantity is mobility ($cm^2\ V^{-1}\ s^{-1}$)

and E is the electric field (Vcm^{-1}). The connection to macroscopic observables is cast into the Ohm law:

$$\overline{J} = qn\overline{\overline{\mu}} \cdot \overline{E} = \overline{\overline{\sigma}} \cdot \overline{E} \tag{8.19}$$

$\overline{\overline{\mu}}$ is the mobility tensor and $\overline{\overline{\sigma}}$ the conductivity tensor. From now on we drop the tensor notation, just keeping in mind that crystalline structures may show anisotropic transport.

The Drude model (See Box 8.2), developed in the beginning of the nineteenth century, just after the discovery of the electron as the elementary negative charge, suggests a simple microscopic relation for the conductivity:

$$\sigma = \frac{ne^2\tau}{m^*} \tag{8.20}$$

where "e" is the electronic charge, τ the scattering or relaxation time, and m^* the effective mass (Drude first proposed his model using the free electron mass). Drude model is incorrect, but it catches the essentials of transport in a solid. Drude had in mind the pinball problem, and assumed that charge carriers (electrons in metals) were moving freely under the electric field action. Owing to the presence of ions, however, those particles were also suffering collisions and sudden changes in momentum. Measuring σ and introducing the material parameters n and m^*, the scattering time τ can be estimated experimentally. It turns out to be in the range of 10 fs for many known metals. Assuming classical statistics for the electron gas to estimate the electron velocity, Drude obtained fairly good agreement with his model

Box 8.2: The Drude Model

The Drude model describes transport in a solid (metal) as occurring in a gas of free electrons. For a single electron, the motion under an external field is described by the equation of motion:

$$m_e\frac{\mathrm{d}v}{\mathrm{d}t} = -eE_0 - \frac{m_e v}{\tau} \tag{8.21}$$

where a damping force is introduced. This damping is caused, according to Drude, by collisions among the carriers and the fixed ions in the lattice, as a ball moving down a pinboard. The speed of the electron "k" at time t is given by $\overline{v}_k = \overline{v}_{uk} - \frac{e\overline{E}_0}{m_e}t_k$, where t_k is the time after the last collision, and v_{uk} is the speed out of the collision. Taking the average of v_k over all N electrons in the metal,

$$\overline{v}_\mathrm{D} = \frac{1}{N}\sum_k\left(\overline{v}_{uk} - \frac{e\overline{E}_0}{m_e}t_k\right) = \frac{1}{N}\sum_k\frac{e\overline{E}_0}{m_e}t_k = \frac{e\overline{E}_0}{m_e}\tau \tag{8.22}$$

because v_{uk}, being random, averages to zero. This is the same result that one gets by directly solving Eq. 8.21 for a stationary case. The average time τ is the "relaxation time" that can be assimilated to an average time between collisions.

because the calculated intercollision path was of the same order of magnitude of the inter-ion distance (lattice constant).

Once the correct statistic was introduced with the development of quantum mechanics (Fermi-Dirac in this case), it became clear that collisions could not take place at ions in the lattice, for the expected mean free path was an order of magnitude larger than the lattice constant. The refined theory considers scattering with phonons and describes the whole process within the wave picture for propagation. Accordingly, electrons in a lattice are Bloch waves that do not scatter onto the fixed ions. Yet the intuitive idea that electrons bounce onto ions remains a very educational and vivid one.

Putting together Ohm and Drude equations we get $\mu = \frac{e\tau}{m^*}$ for the mobility. It suggests that low mobility might stem from large effective mass (flat dispersion curve) and short scattering time (which implies strong interaction of the carrier with the medium or, in classical terms, high viscosity against motion). Even if we are using over simplified expressions, we just obtained very fundamental statements on transport, which remains true for the real phenomena: any scattering event (due to impurities, defects, phonons, and other quasiparticles) reduces the ability of a material to transport charges. In addition, effective mass is a concept of band theory, but what we said is general, classical, and does not require extended states. One can simply replace the effective mass with the mass of the carrier, be it an ion, particle, or molecule.

Diffusion is the motion that occurs because of a gradient of concentration (relevant in DSSC photovoltaic cells). The resulting current will bring the system to equilibrium, at homogeneous distribution, if the cause of dishomogeneity is not permanent. The phenomenological description of diffusion is given by the Fick equation

$$\bar{J} = -qD \cdot \overline{\nabla} n \tag{8.23}$$

where D is the diffusion constant ($\mathrm{cm^2\,s^{-1}}$). This is a second-rank tensor but reduces to a scalar quantity for isotropic materials. The important difference here with respect to drift current is that there is no driving electric field. The driving force is the concentration gradient. The mean free path is connected to D and τ (this could be the lifetime of the carrier) by the simple relation $\langle x^2 \rangle = nD\tau$, with $n = 2, 4, 6$ for 1D, 2D, and 3D spaces.

By adding the conservation of the particle number in the diffusion process the second Fick law in 1D is obtained:

$$\frac{\partial n(x,t)}{\partial t} = D\frac{\partial^2 n}{\partial x^2} \tag{8.24}$$

With the initial condition that all particles at $t = 0$ are in $x = 0$, the solution is

$$n(x,t) = \frac{N}{\sqrt{4\pi Dt}} \exp\left(-\frac{x^2}{4Dt}\right) \tag{8.25}$$

which describes a Gaussian around $x = 0$, broadening with time.

In the presence of an electric field, both phenomena, drift and diffusion, take place. The current density is then given by

$$\bar{J} = qn\bar{v}_\text{D} = qn\bar{\bar{\mu}} \cdot \bar{E} - qD \cdot \bar{\nabla} n \tag{8.26}$$

and the population concentration, in 1D, by

$$\frac{\partial n(x,t)}{\partial t} = -v_\text{d}\frac{\partial n}{\partial x} + D\frac{\partial^2 n}{\partial x^2} \tag{8.27}$$

whose solution is a time broadening, propagating Gaussian $n(x,t) = \frac{N}{\sqrt{4\pi Dt}} \exp\left(-\frac{(x-v_\text{d}t)^2}{4Dt}\right)$.

Finally, the two processes are connected by the Einstein–Smoluchowski equation $\mu = \frac{qD}{k_\text{B}T}$, where q is the carrier charge. This can be very useful when experiments can access D, but one likes to know μ, or vice versa. It also specifies the common ground of the two phenomena. The Einstein–Smoluchowski equation is obtained by postulating the equilibrium between drift and diffusion that cancels out current.

8.5.1 Experiments to Measure Mobility

8.5.1.1 Time of Flight (ToF)

The most direct way to measure mobility is to perform a time-of-flight (ToF) experiment (Figure 8.11). The sample is sandwiched between two electrodes, one semitransparent (for instance, gold/ITO and semitransparent Al). A pulsed optical excitation injects a sheet of carriers (a bunch of charge particles well confined in space) near the semitransparent electrode. The experiment measures the time it takes for the carrier wavepacket to reach the opposite electrode. According to the applied bias, positive or negative carriers drift to the bottom electrode. If the wavepacket is Gaussian and it remains reasonably so, mobility is obtained from a simple kinetic analysis: $\mu = \frac{v_\text{D}}{E} = \frac{d^2}{V\tau}$, where τ is the transit time, d the distance between the injection point and the collection point, and V the applied voltage (to provide field $E = V/d$). In the current mode, when the circuit RC $\ll \tau$, the measured current shows a plateau in time till an inflection point and then decays off. The inflection point defines the *transit time* τ. A current plateau for $t < \tau$ indicates that charge carrier generation is completed by $t \ll \tau$ and no carriers are lost during their motion. In addition, this suggests that carrier velocity is time independent. Under such circumstances, only diffusion broadens the wavepacket.

In a time $\tau = \frac{d^2}{\mu V}$, the average diffusion length is $\Delta x = \sqrt{2D\tau} = \sqrt{\frac{2Dd^2}{\mu V}} = d\sqrt{\frac{2kT}{eV}}$, where we assumed that the Einstein–Smoluchowski equation is valid. The spread in arrival time $\Delta\tau = \frac{\Delta x}{v_\text{D}} = \frac{d}{\mu V} d\sqrt{\frac{2kT}{eV}} = \tau\sqrt{\frac{2kT}{eV}}$ is fairly small under typical experimental values ($V = 500$ V, room temperature, $\Delta\tau/\tau \approx 10^{-2}$).

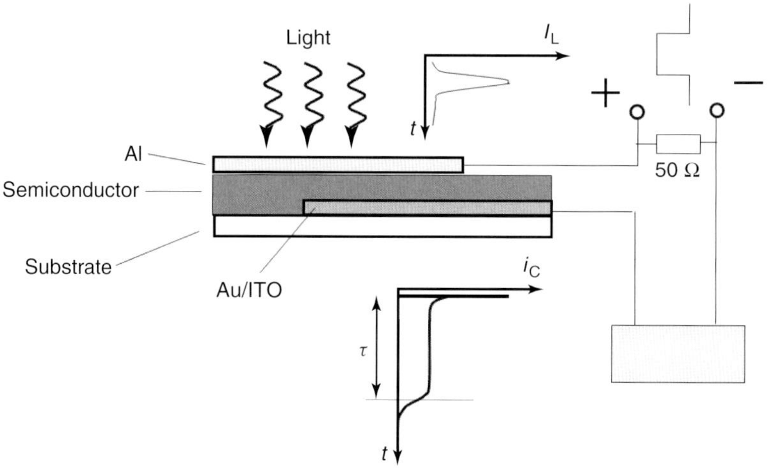

Figure 8.11 Layout of the time-of-flight experiment. I_L is the light intensity; i_C is the current. The dashed line in the current plot shows the inflection point to determine the transit time τ.

Owing to disorder there can be a large distribution of transit times. At some point it can happen that the initial wavepacket does not really propagate but just gets stretched toward the collecting electrode. When this happens, the uncertainty of the measurement is increased, and the inflection point occurs on a broad tail. In the extreme condition when some carriers already reached the collection, others still have to start. The lack of a well-defined transit time is a trouble for this experiment and calls for a statistical interpretation. In this case, transport is dispersive, meaning that there is a distribution of hopping times for the carriers. Some are very quick, others practically immobile. This behavior is a clear manifestation of disorder, and it is typical of organic amorphous semiconductors.

8.5.1.2 Photoconductivity

Photoconductivity is the effect of enhanced conductivity in a semiconductor under illumination (Figure 8.12). Usually this experiment brings information on the charge generation process rather than on transport, but the two are intimately linked here. Light illuminates the whole specimen and the charge carrier profile is less crucial with respect to ToF. The measured current contains the product of mobility, carrier lifetime, light intensity, and applied field. At low and high light intensities the population rate is described as follows:

$$\frac{\mathrm{d}n_L(x)}{\mathrm{d}t} = \eta\alpha I\mathrm{e}^{-\alpha x} - \frac{n(x)}{\tau} \tag{8.28}$$

$$\frac{\mathrm{d}n_H(x)}{\mathrm{d}t} = \eta\alpha I\mathrm{e}^{-\alpha x} - \beta n^2(x) \tag{8.29}$$

where I is the light intensity in photons per unit area and α is the absorption coefficient such that $\mathrm{d}I = -\alpha I\mathrm{e}^{-\alpha x} = -\alpha I(x)$ is the number of photons absorbed

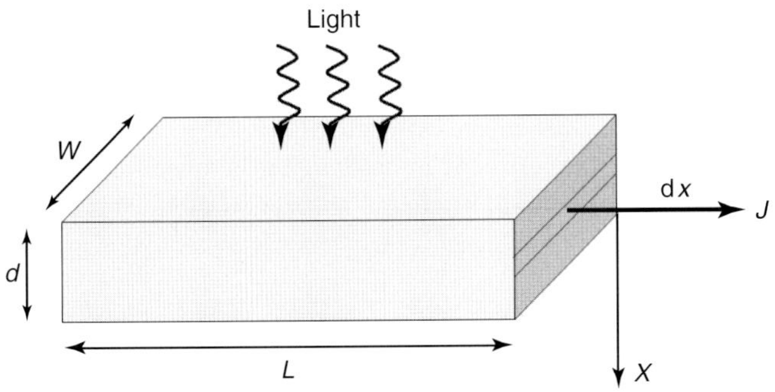

Figure 8.12 The photoconductive element in the photoconductivity experiment. Collecting electrodes are at the two sides of the slab.

at depth x. η is the charge generation quantum efficiency (number of free carriers generated per absorbed photons, $\eta = \Phi\Omega$, where Φ is the quantum yield of charge generation and $\Phi(E)$ the escape probability from geminate recombination, usually a function of the electric field). The monomolecular lifetime is τ, and the bimolecular recombination constant is β. The two equations are simplified, and assume only monomolecular recombination at low light intensity and only bimolecular recombination at high light intensity. The collected current in the geometry of the experiment is

$$i_C = \int_0^t qn(x)\mu E \, w\mathrm{d}x \tag{8.30}$$

where E is the applied electric field and w the sample width. At steady state the charge carrier populations are

$$n_L(x) = \eta \, \tau\alpha I \mathrm{e}^{-\alpha x} \tag{8.31}$$

$$n_H(x) = \left(\frac{\eta\alpha}{\beta}\right)^{1/2} \sqrt{I}\mathrm{e}^{-\frac{\alpha}{2}x} \tag{8.32}$$

which leads to

$$i_C = qw\eta\mu\tau I \left(1 - \mathrm{e}^{-\alpha d}\right) E \tag{8.33}$$

$$i_C = 2qw\left(\frac{\eta I}{\alpha\beta}\right)^{1/2} \mu\left(1 - \mathrm{e}^{-\frac{\alpha}{2}d}\right) E \tag{8.34}$$

for monomolecular and bimolecular recombinations, respectively. If charge carriers in the sample are fully collected at electrodes, then their lifetime is $\tau = \frac{L}{\mu E}$, and the low-intensity current, $i_C = qw\,\eta L I \left(1 - \mathrm{e}^{-\alpha d}\right)$ depends only on the number of absorbed photons $I\left(1 - \mathrm{e}^{-\alpha d}\right)$ and the quantum generation efficiency, regardless how fast or slow the carriers reach the electrode. Once η is known, a measure of the photocurrent after a short-pulse excitation ($t_{\text{Light}} < \tau$) allows determination of the carrier velocity and thus mobility. In general, this is not the case, and photoconductivity experiments alone provide information on the $\eta\mu\tau$ product. Combining photoconductivity with optical absorption measurements, specifically charge-induced absorption, is a way to circumvent this problem and separate generation from transport. This experiment is challenging because one should know precisely enough the cross section of charge absorption (doublet–doublet transition).

8.5.1.3 Field Effect Transistor (FET)

A field effect transistor (FET) operates as a capacitor, whereby one plate is a conducting channel between two ohmic contacts, the source, and the drain electrodes. The density of charge carriers in the channel is modulated by the voltage applied to the second plate of the capacitor, the gate electrode. In this experiment, we can consider two regimes, the ohmic or linear regime and the saturation regime (Figure 8.13). In the linear regime $V_{DS} \ll V_G - V_T$, where V_G is the gate potential and V_T is the device threshold voltage. The source-drain current (I_{SD}) is linear with applied potential V_{DS} according to

$$I_{SD} = \frac{W}{L}\mu_L C\left(V_G - V_T)\right) V_{DS} \tag{8.35}$$

where C is the capacitance of the gate dielectric per unit area.

In the saturation regime, the transistor charge distribution has the characteristic profile marked by a point of "pinch-off" where the density of carriers becomes zero. At threshold $V_{DS} = V_G - V_T$ and

$$I_{SD} = \frac{W}{2L}\mu_S C(V_G - V_T)^2 \tag{8.36}$$

As a consequence of $V_{DS} > (V_G - V_T) I_{SD}$ is independent from V_{DS}.

Transistors are characterized by the "output characteristic," plotting source-drain current versus source-drain voltage, and "transfer characteristic," plotting source-drain current versus gate voltage.

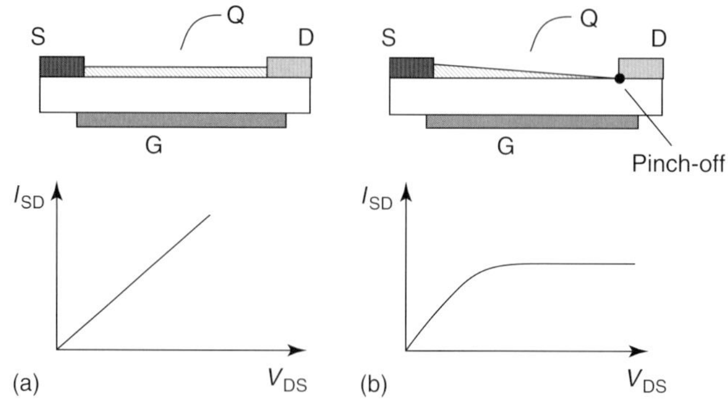

Figure 8.13 (a) FET in linear regime. Dashed area is the charge profile. (b) FET in saturation.

A few observations on this measurement are

1) The channel of the FET is very thin, corresponding to few molecular layers. This implies a peculiar morphology, characteristic of the first deposited layers, which usually change on making thicker samples. For this reason, the mobility measured in a transistor may be different from that measured in a bulk sample, with different molecular packing, by other techniques.
2) The two mobilities, at linear, μ_L, and saturation, μ_S, regimes are generally different. This may stem from a different state of filling of the trap in transporting layer because of a different carrier density profile.
3) The electric field distribution in a ToF or FET experiment is different, contributing to a different value of mobility.

In general $\mu_{FET} > \mu_{ToF}$.

8.5.1.4 **Pump Probe**

There are at least two experiments of pump probe that can lead to an estimate of charge carrier mobility. The common requirements are (A) pump-induced charge generation in the semiconductor and (B) the possibility to access spectroscopically the charge carriers by measuring charge induced absorption. Let us assume that both holds true. Then one can measure bimolecular recombination of the charge carriers. In order to exclude monomolecular (geminate) decay, pump intensity dependence of the recombination kinetics should be clearly observed. Assuming this is true, the charge population is an observable in the pump probe experiment. The rate equation for charge carrier population is

$$\frac{\mathrm{d}n_D}{\mathrm{d}t} = -k(n_D)n_D \tag{8.37}$$

where n_D is the photoinduced doublet (charge carrier with spin = 1/2 multiplicity) population, generation is assumed to be instantaneous ($n_D(0) \neq 0$), and the decay rate k is population dependent, $k = \gamma n_D$, giving the expected square dependence of the decay rate. The bimolecular constant γ can be measured by fitting the pump probe time trace, which is monitoring charge population, for instance, looking at the peak of charge-induced absorption.

From the bimolecular recombination coefficient one can obtain an estimate of mobility according to the Langevin model, as explained in the following. We assume that charge carriers are statistically independent and the mean free path (λ) is smaller than the Coulomb capture radius (r_C), $\lambda < r_C$. The Coulomb capture radius in organic semiconductor is $r_C = \frac{e^2}{4\pi\varepsilon k_B T} \approx 20$ nm, while the typical mean free path is $l \approx 0.1$–1 nm, so this condition is usually fulfilled. Under all these circumstances, Langevin recombination, as it was proposed in 1903, holds true, and recombination is a random process with bimolecular kinetics. Within Langevin recombination, charges move under the mutual Coulomb attraction till coalescence (Figure 8.14). Let us assume one of the charges (electron) is fixed at the center of a sphere of radius r_C. The hole density current (hole flux) through the sphere is

$$J = n_h e\mu \frac{e}{4\pi\varepsilon r_C} \tag{8.38}$$

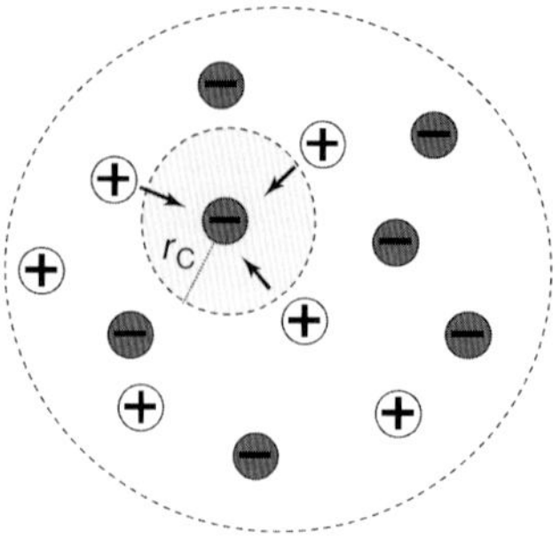

Figure 8.14 Within the pump spot, a population of statistically independent charge carriers is produced (dashed circle). The dashed small circle shows the Coulomb capture sphere through which a hole current leads to recombination.

and the corresponding current equals to the recombination rate times the electron charge: $i = j \cdot 4\pi r_C^2 = n_h \mu \frac{e^2}{\varepsilon} = e\frac{dn_h}{dt} = e\gamma n_h$. This leads to the final equation $\gamma = \frac{e}{\varepsilon}\mu$, which links the bimolecular recombination rate, measured by pump probe, with the mobility.

This analysis depends on a number of assumptions, each of them to be carefully validated. However, it also has a few very interesting and peculiar characteristics. (i) It provides mobility values for very short time after excitation, typically subpicoseconds up to nanoseconds. At this time scale, one gets information on hot carriers, before thermalization has taken place. (ii) Time-dependent mobility can be obtained. (iii) Because of the short time scale, the measured value is relative to a very small space, that is, it is a local probe of transport properties. (iv) With respect to previous experiments, the one we present now is electrode free, and because it corresponds to small traveled distances for the carrier, it is also free from border regions between crystallites or other defects. This technique can measure crystalline mobility in polycrystalline samples if the crystallites are larger than the probe size. In a standard experiment the probe area is 100 μm, but using confocal microscopy this value can be considerably reduced to submicrometers.

A second experiment is photoinduced anisotropy decay (Figure 8.15). Besides requirements (A) and (B) above, the carrier optical transition should be linearly polarized, and both pump and probe light pulses should be linearly polarized. Then, on excitation, the initial photoinduced absorption due to charge states will show anisotropy, as described for excitons. The loss of anisotropy is associated with charge carrier diffusion. With a proper guess on the morphology and the path traveled in space, ΔX, one can infer the diffusion constant $D \approx \frac{\Delta x^2}{\tau}$, where τ is the anisotropy decay constant (see the pump probe section). From this and the Einstein–Smoluchowski relation, mobility can finally be obtained.

Transient grating is a similar experiment that could be adopted for transport characterization, (Figure 8.16). Assumed previous conditions are still true. In this experiment, the initial two-pulse interference creates a periodic pattern in the sample space, with sheets of carriers separated by unexcited regions. The measured

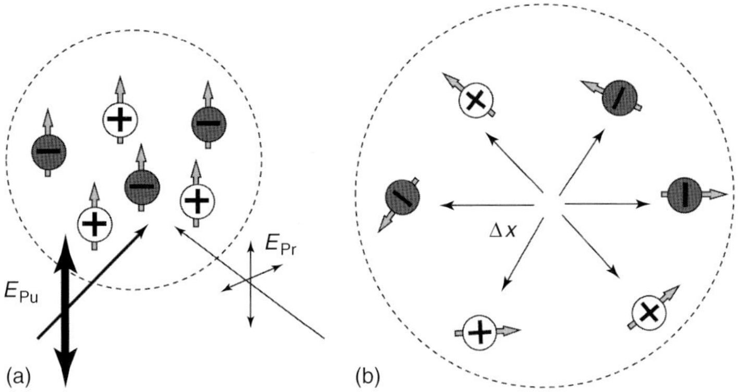

Figure 8.15 (a) The initial photoinduced population is linear polarized (oriented) parallel to the pump field. (b) After diffusion, the anisotropy is lost. Note that in true experiment on an isotropic sample the initial orientation distribution is not 100% as shown, but it follows a cosine square dependence around the pump field vector. A guess on the average distance traveled, Δx, allows to work out a diffusion constant.

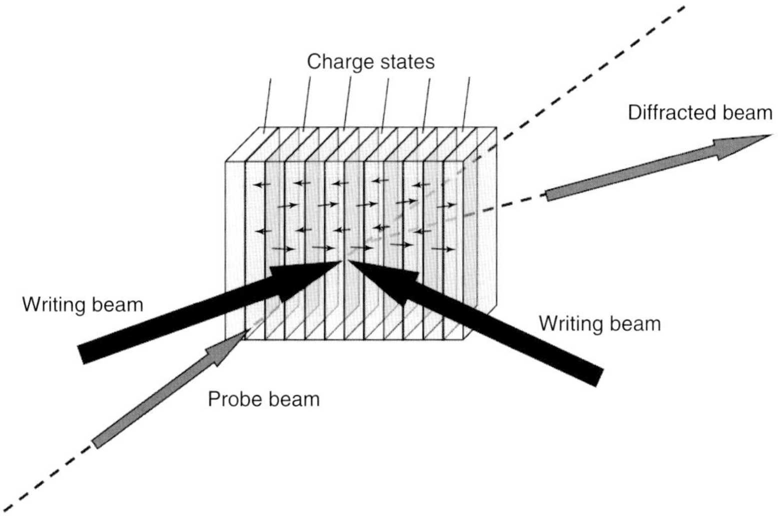

Figure 8.16 The initial concentration grating contains layers of charge states separated by unexcited volumes. Owing to diffusion (the little black arrows) this modulation will be washed out and the amplitude of the diffracted beam will consistently decay off.

quantity is the diffraction efficiency of the third, probing beam, which depends on the contrast ratio of the photoinduced grating. On diffusion of charge carriers the grating gets washed out. Because one can distinguish between recombination and diffusion, by measuring separately the population decay, this experiment also provides information on carrier diffusion, and thus mobility.

Finally, we briefly mention long wavelength probing. Free carriers might not be associated with optical transitions (or near-IR transitions), because they have a quasi continuum spectrum of electronic levels. In this case, terahertz or microwave probing can be useful for detecting the charge carriers. In addition, in this range of electromagnetic interactions the response function is directly the conductivity of the sample, that is, the imaginary part of the dielectric function. In such experiments, the pump generates a population of carriers and the terahertz/μ-wave probe measures the transient conductivity. The pump may be an ionizing radiation, such as an electron beam. In this case, the excitation pulse delivers electrons of few megaelectron volts energy and it lasts a few nanoseconds. The probe is radiation of 30 GHx frequency and 100 mW power, with a corresponding field in the order of 100 Vcm^{-1}. This is several orders of magnitude smaller than that used in ToF. The experiment can still probe local properties and also virtually access isolated large molecules or polymer chains. In a quarter period of μ-wave, 10 ps, the diffusion length $\lambda = \sqrt{2Dt}$ of a charge with 1 $\mathrm{cm^2\ V^{-1}\ s^{-1}}$ mobility, that is, diffusion constant $D = 2.5 \times 10^{-2}\ \mathrm{cm^2\ s^{-1}}$, is 7 nm. The radiation-induced, transient conductivity is $\Delta\sigma = en\,(\mu_+ + \mu_-)$, where n is the concentration of the generated carriers. This is a critical parameter for the actual evaluation of mobility, which can introduce uncertainty in the measurement because the ionization efficiency may not be known precisely. Besides using test samples one could think of developing an optical/μ-wave technique in which optical probing allows evaluating the charge carrier concentration after ionization. This might work when charge carriers are actually ions with well-defined electronic transition in the visible/near-IR region. The drawbacks are that the transition cross section for charge-induced absorption is also not known precisely and optical measurements do not distinguish between trapped and free charge carriers. In addition, both positive and negative charge carriers contribute necessarily to the experimental response.

Better time resolution can be obtained using terahertz probing (typically frequency of 1 THz and time duration of about 1 ps) and optical pumping with 100 fs pulses. Again, providing efficiency of charge photogeneration is known, the attenuation of the terahertz pulse transmitted by the excited sample can be used to infer a value of mobility through imaginary dielectric function. There are however pitfalls in the data analysis, for displacement currents might add up to conductive currents.

8.5.1.5 **Diode**

In a single layer diode structure the semiconductor is sandwiched between two electrodes. We assume that, the semiconductor does not have free carriers. This is true for any semiconductor at $T = 0$ or for organic semiconductors, which have low conductivity. In this condition and under bias, carriers are supplied by injection at the electrodes. At low voltage bias, V, the observed current through the device is ohmic, that is, linear with V. In this regime the electrode supplies the carriers, usually of one sign, depending on the level alignment at interface. At some point, even if the electrode is able to supply an infinite number of charge carriers, the current is limited by its own space charge, which eventually reduces the electric field at the injecting electrode to zero. It happens approximately when the charge carrier concentration equals the capacitor charge $Q = \varepsilon \frac{A}{d} V$. This charge transits through the device in a time $\tau = \frac{d^2}{\mu V}$ so that the space charge limited current (SCLC) is $J = \frac{Q}{A \cdot \tau} \approx \varepsilon \mu \frac{V^2}{d^3}$. This simple calculation ignores the inhomogeneous distribution of the electric field in the bulk of the sample. The correct value of the stationary SCLC derived by using the Poisson equation and continuity equation is, however, almost the same, $J = \frac{9}{8} \varepsilon \mu \frac{V^2}{d^3}$, which is known as *Child's law*. So an I–V characteristic in the SCLC regime can be used to obtain mobility.

Reality is much more complicated than this, because of phenomena at the injecting contact and trapping of carriers. One finds a broad range of experimental values of the kind $J \propto \frac{V^{l+1}}{d^l}$ with $l > 1$. Handling this situation requires some experience in transport measurements and a more extensive investigation of the phenomenon, for instance, varying systematically the sample thickness. This might, however, introduce other variables to the problem, because bulk morphology depends on thickness and the mobility from the morphology.

8.5.1.6 **Optical Transitions**

Optical transitions in many materials can be changed by applying static electric fields, which affect the electronic states. When this occurs on confined states, it is usually called *STARK effect*. In extended states the phenomenon is known as *Franz–Keldsyh effect*. Electroabsorption and electroreflectance are common experimental geometries. Both can be exploited for measuring charge carrier mobility. The technique was first reported to measure nonequilibrium carrier transport in inorganic amorphous semiconductors by Shank and coworkers [5, 6]. Here we briefly review the experiment as done in organic semiconductors of very low mobility, where STARK effect is responsible for the modulation of ground-state transmission in the presence of a field. The material under study is sandwiched between two electrodes, one semitransparent (Figure 8.17). A static electric field is applied to the diode structure in reverse bias, to avoid charge injection. A probe pulse is transmitted through the sample (or back reflected by one of the electrodes) and used to measure transmission, T. In the presence of a static electric field such transmission includes the field-induced effect, $T(F) = T_0 + \Delta T(F)$, where F is the applied field. A resonant pump pulse excites the material and generates charge carriers. This process might be direct or occur on dissociation of the primary neutral excitations. In any case charge carriers are generated in the sample, and they start to drift under the action of the field. Charge separation will build up a polarization that quenches the field. Accordingly, the field-induced transmission of the probe, $\Delta T(F)$, will be reduced, restoring the neutral value T_0. Because the screening effect depends on charge separation, and this grows in time upon charge carrier drifting, the technique offers a handle for measuring mobility.

Starting from the effective linear pump probe equation

$$\frac{\Delta T}{T} \cong -\sum_{i,j} \sigma_{ij}(\omega) \Delta N_j(I_{\mathrm{Pu}}, \tau) D \qquad (8.39)$$

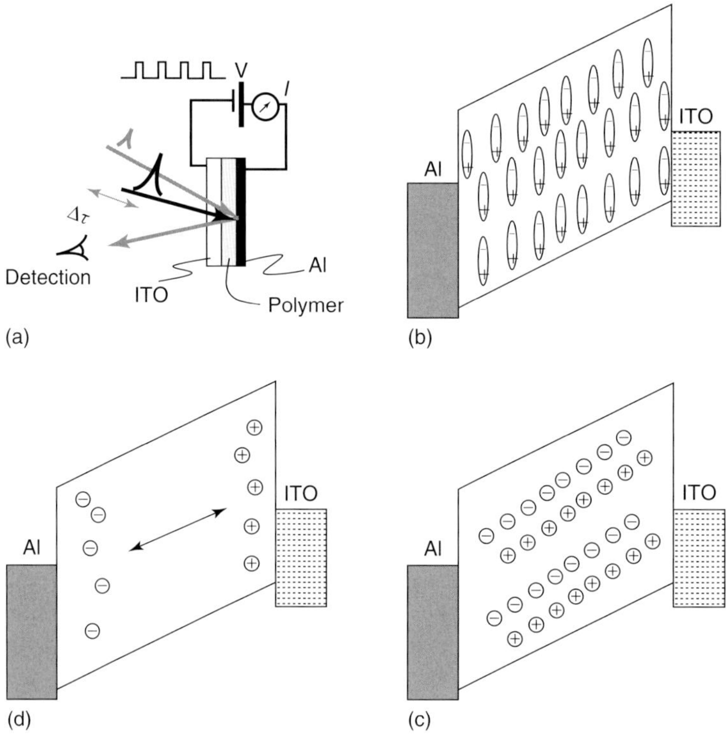

Figure 8.17 (a) The simplified experimental layout. The initial (b) excitation, (c) ionization, (d) and charge separation leading to field screening.

The effect of the electric field on the pump probe signal can be described by

$$\Delta^2 T/T = -d\sum_{i,j} \Delta\sigma_{ij}(\omega, F)\Delta N_J \otimes f_\mathrm{p} - d\sum_{i,j} \sigma_{ij}(\omega)\Delta^2 N_J(F) \otimes f_\mathrm{p} - \sum_{i} \Delta\sigma_{i0}(F(t))N_0 \tag{8.40}$$

The first term regards a change in the cross section of excited-state transitions (pump induced), the second regards the change in the pump-induced population, and the third regards the change in the ground-state cross section. Here we are concerned with the third one. We explicitly wrote the field $F = F(t)$ to stress the effect we are looking at: a time-dependent reduction in the internal electric field due to the photoinduced charge screening.

The effective field acting on the molecules in the sample is

$$F_\mathrm{eff}(t) = F_\mathrm{app} - \frac{P(t)}{\varepsilon_0} \tag{8.41}$$

where the time-dependent polarization is due to the presence of the photoexcited charge carriers, separated by a distance $R(t)$, which grows in time:

$$\overline{P} = \sum_i e\overline{r}_i(t) = Ne\langle \overline{R}(t) \rangle \tag{8.42}$$

The measured effect on the probe transmission is quadratic in the effective field and will be reduced in time according to the increasing polarization

$$\frac{\Delta T(t)}{T} \approx \frac{1}{2} \Delta p \frac{\partial \alpha}{\partial E} F_{\text{eff}}(t)^2 \tag{8.43}$$

This equation allows working out the effective field, the polarization, and finally the average carrier distance (dipole length). Δp is the polarizability change between the ground and excited states associated with the optical transition under study. Once $R(t)$ is known, the drift velocity and finally the time-dependent mobility can be obtained.

$$v_{\text{drift}} = \frac{d < R(t) >}{dt} \tag{8.44}$$

$$\mu(t) = \frac{v_{\text{drift}}(t)}{F_{\text{eff}}(t)} \tag{8.45}$$

This technique is intriguing, because it allows obtaining mobility in the subpicoseconds time domain, without current extraction. It is suitable for measuring hot carrier transport and can provide local information, for instance, crystalline mobility in a polycrystalline sample. The disadvantage is the number of strict requirements that are needed. The pump excitation should generate charge carriers; charge generation should be instantaneous; the wavelength at which evolution of the electroabsorption is measured should be free from other field-assisted contributions; and an estimate on the number of photogenerated charges should be available. When all these requirements are fulfilled the experiment is quite straightforward. If not, one can still extract information with a careful analysis of the signal in a suitable spectral range, provided a good assignment of all the bands in the transient transmission spectrum is available.

8.5.2 Extended versus Localized States

The two extreme situations in a solid material are *coherent* transport through extended states or *incoherent* transport among a distribution of localized states; see Chapter 4 for a more detailed discussion. Coherent transport is the wavelike propagation of a quasiparticle, which is described by the semiclassical wavepacket state. Wavepackets are coherent superpositions of extended Bloch states, which propagate under the dispersion law in k-space, according to the group velocity $v_G = \frac{1}{\hbar}\frac{\partial E}{\partial k}$. The k-vector is a good quantum number that describes quasi-momentum of the propagating quasiparticle. The notion of transport as displacement of a quantity, here electrical charge, is provided by the quasi-classical state that allows defining two observables: position and momentum. Scattering events change the

k-vector. When this occurs too often for wave propagation to be established, that is, the k-vector is no more well defined, coherent transport is lost. This is associated with a transition from extended states $\Psi_k(r) \propto e^{ikr}$ to localized states $\Psi_k(r) \propto e^{-r/a}$, where "$a$" is a characteristic distance typically of the order of the orbital wavefunction. When this happens, transport becomes incoherent and it is described by hopping: a finite probability for the charge particle to jump from one site to another in the solid. Typically, electron transfer occurs as a result of redox reactions between nearby molecules in a fixed geometry. In both cases, of coherent and incoherent transport, energy bands describe the electronic structure. In the coherent case, however, (i) bands develop in k-space, (ii) bands are associated with fully delocalized states in real space, and (iii) the real space is a periodic lattice of well-ordered sites. In the incoherent case (i) the energy bands develop in real space, (ii) each site in the real space has energy and coupling different from its neighboring sites. and (iii) the solid is usually amorphous or highly disordered. The width of the coherent band in k-space depends on the strength of the interatomic or intermolecular interaction, while the width of the incoherent band in real space depends on disorder (Figure 8.18).

Transport modeling is based on master equations ruling the occupation of sites and the coupling between sites in the phase space (coherent) or real space (incoherent). Often modeling relies on statistical methods such as the Montecarlo method to describe the evolution of an initial distribution of carriers in space.

Here we are concerned with organic semiconductors and we will describe incoherent transport in some more detail. The fundamental event is hopping from one site to the other. Many hopping events lead to macroscopic transport, but the single event may not be driven by external forces in a deterministic way, that is, the jump may occur against the field.

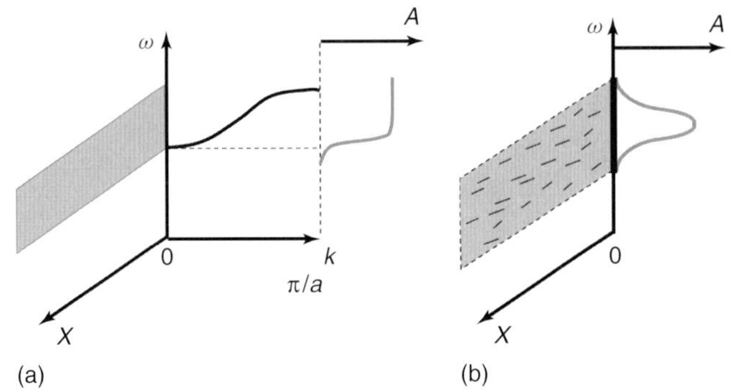

Figure 8.18 (a) Conduction band in a crystal. In real space (X) states are fully delocalized. In k-space, energy dispersion leads to a band. A is the joint density of states, or absorption. (b) Disorder-induced band, containing a distribution of site energies in real space.

One famous model for hopping probability between two sites in a random array of molecules is the Miller–Abrahams model. Here we give a brief survey of the concepts involved.

The hopping rate between site i and j according to Miller–Abrahams is given by

$$\nu_{ij} = \nu_0\, e^{-2\gamma R_{ij}} \quad \text{for } \varepsilon_j < \varepsilon_I \tag{8.46}$$

$$\nu_{ij} = \nu_0\, e^{-2\gamma R_{ij}} e^{\left(-\frac{\varepsilon_j - \varepsilon_i}{k_B T}\right)} \quad \text{for } \varepsilon_j > \varepsilon_I \tag{8.47}$$

where ν_0 is an attempt frequency that depends on electron–phonon coupling and the phonon DOS, R_{ij} is the distance between site i and j, γ is the overlap factor, ε_i and ε_j are site energies, and the first exponential term is a T-dependent tunneling probability, while the second exponent is the Boltzman factor, which describes the thermal activated rate accounting for the T dependence of the phonon DOS. Drift under an electric field requires an additional term $-e\overline{R}_{ij} \times \overline{F}$ in the energetic balance.

Alternatively, the Marcus hopping rate is given by

$$\nu_{ij} = \frac{t^2}{\hbar}\left(\frac{\pi}{\lambda k_B T}\right)^{1/2} \exp\left[-\frac{(\lambda + \varepsilon_j - \varepsilon_i)}{4\lambda k_B T}\right] \tag{8.48}$$

Both expressions can describe diagonal disorder, i.e. the distribution in site energy (ε_i) and off-diagonal disorder, i.e. the distribution in intersite coupling strength ($t,\ R_{ij},\ \gamma(R_{ij})$). The Marcus theory includes the inverted region that is totally absent in the Miller–Abrahams model. The Miller–Abrahams model is used when the electron–phonon coupling is weak and temperature is low. In this regime $(\varepsilon_{i–}\varepsilon_j) < h\nu_V$, and processes are single phonon. Marcus hopping rate describes strong electron–phonon coupling and high-temperature (multiphonon) processes.

Montecarlo simulations using the rate formula as above on a lattice of NxN sites allows extracting mobility by fitting the simulated behavior of a population of carriers or, alternatively, the outcomes of many carrier evolution paths. The master equation describing the occupation probability f_k of a site in the lattice is

$$\frac{df_i}{dt} = \sum_{j \neq i} \nu_{ji} f_j - f_i \sum_{j \neq i} \nu_{ij} \tag{8.49}$$

where ν_{ij} is the hopping rate as specified above. Recombination (disappearing of the site occupation) is not considered in Eq. (8.49), but it could be easily included. Under certain conditions, Eq. (8.49) can be solved even analytically. An example is the perfect crystal in which $\nu = \nu_{ij}$ and $a = R_{ij}$ and nearest neighbor hopping is dominating. The approach is then equivalent to the continuity equation for carrier density with drift and diffusion terms. The obtained diffusion constant and mobility are $D = \nu a^2$ and $\mu = e\nu a^2/k_B T$, which obeys the Einstein–Smoluckowski relationship. By numerical solution, which is similar to carrying out an idealized experiment, transport can be described in more complex situations. A population of carriers distributed according to some initial condition on the check board of NxN sites will be propagated according to the master equation, essentially "flipping a coin" at each hop. This will provide average trajectories of the carriers that can be further analyzed as real data from an experiment. Mobility is one of the properties that can be estimated, as speed per unit of electric field, others are diffusion, free mean path, and so on. For instance, one can get temperature-dependent mobility by repeating the simulation for different temperatures. Using Miller–Abrahams model, Baessler obtained $\mu(T) = \mu_0 e^{\left[-\left(\frac{T_0}{T}\right)^2\right]}$ where $T_0 = \frac{2\sigma}{3k_B T}$ describes the *disorder-induced* distribution of states, with Gaussian bandwidth σ. A value of $\sigma/K_B T$ between 1 and 10 appears reasonable from data simulation. Similarly, by changing the electric field one finds that for very high fields, $eFa \approx \sigma$ disorder effects tend to vanish, v is constant, and $\mu \sim F^{-1}$, while for intermediate fields the Poole–Frenkel behavior $ln\ \mu \sim F^{-1/2}$ is found. The empirical expression for mobility, including the Poole–Frenklel expression for the field dependence, is $\mu_{PF}(T) = \mu(T) e^{\gamma(T)\sqrt{F}}$. One sees from this expression the difficulty in using such models, because of the number of unknown parameters in the formulation. Here, for instance, the temperature dependence of the pre-factor γ will end up in the temperature dependence of the exponent in $\mu(T) = \mu_0 e^{\left[-\left(\frac{T_0}{T}\right)^2\right]}$, masking the disorder-induced behavior.

In the polaron model, transport is limited by a barrier that slows down the carrier according to $\mu(T) = \mu_0 e^{\left[-\frac{\Delta}{K_B T}\right]}$, where Δ is related to the polaron binding energy.

Finally, because transport is concerned with state occupancy, a carrier density dependence is also expected, as deep or shallow traps get filled, changing the distribution of available states, their energies, and coupling rates.

Many phenomenological expressions can be obtained for describing mobility, but a general solution does not exist. The process is intrinsically governed by disorder, and the best description is at a statistical level as given by a many-particle Montecarlo simulation.

References

1. Marcus, R.A. and Sutin, N. (1985) *Biochimica et Biophysica Acta*, **811**, 265–322.
2. Jortner, J. (1976) *J. Chem. Phys.*, **64**, 4860.
3. Popovic, Z.D. (1984) *Chem. Phys.*, **86**, 311.
4. Noolandi, Z. and Hong, K.M. (1979) *J. Chem. Phys.*, **70**, 3230.
5. Shank, C.V. and Auston, D.H. (1982) *Science*, **215**, 797.
6. Shank, C.V. *et al.* (1981) *Appl. Phys. Lett.*, **38**, 104.

9
Pump Probe and Other Modulation Techniques

Perturbing a system in controlled and reproducible manner is one of the most effective ways to investigate its nature. In spectroscopy, this translates into a large set of techniques where temperature, applied electric field, applied magnetic field, impinging light, and more are modulated. The idea in common to all such techniques is to induce small changes in the sample properties that are measured by using differential detection. As a consequence, derivativelike features are common in modulated difference spectra. The time regime of the modulation can vary from impulsive to quasi steady state. Typically, temperature and magnetic field are slowly modulated, while electric field can reach the gigahertz range and light can cover many decades, from millisecond to femtosecond time domain. In this chapter, we focus on three techniques, namely, electroabsorption, pump probe, and cw-photomodulation. They are all considered in terms of transmission and correlated to photophysics of organic semiconductors.

9.1
Electroabsorption

A static electric field acting as a perturbation can change the absorption spectrum. On a confined molecular or excitonic state the static field can

1) Shift the energy according to the state properties, thus affecting electronic transitions.
2) Ionize the state, leading to lifetime broadening and charge generation.
3) Break the symmetry of the system, relaxing selection rules.

The energy shift effect is the most investigated, and in molecules, it is described by the quadratic Stark effect. The energy of an electronic transition from ground state to excited state, $0 \rightarrow k$, can be classically expanded on the field as

$$\Delta E = E - E(0) = -m_{0k}F - \tfrac{1}{2}p_{0k}F^2 + \cdots \qquad (9.1)$$

where m_{0k} is the change in permanent dipole moment and p_{0k} is the change in polarizability going from 0 to k-state. In the presence of the field, the absorption spectrum shifts according to this change in energy, as depicted in Figure 9.1. In

The Photophysics behind Photovoltaics and Photonics, First Edition. Guglielmo Lanzani.

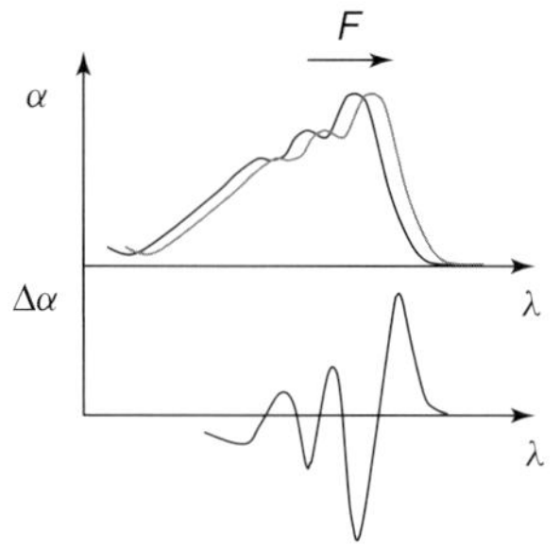

Figure 9.1 Owing to the static electric field F the absorption spectrum shifts to the longer wavelengths. The difference spectrum $\alpha(F) - \alpha(0) = \Delta\alpha$ resembles the first derivative.

electroabsorption, the difference spectrum $\Delta\alpha = \alpha(F) - \alpha(0)$ is detected using a modulation technique. The electric field modulation applied by an external circuit is detected by lock-in amplification. With an applied voltage between 1 and 100 V on molecular films of thickness 100 nm, the applied field is 10^5–10^7 Vcm^{-1}. The absorption shift in energy is small, and the observed difference spectrum resembles a derivative according to

$$\Delta\alpha = \frac{\partial\alpha}{\partial\varepsilon}\Delta\varepsilon + \frac{1}{2}\frac{\partial^2\alpha}{\partial\varepsilon^2}\Delta\varepsilon^2 \tag{9.2}$$

Inserting Eq. (9.1) in (9.2) we obtain

$$\Delta\alpha = -\frac{\partial\alpha}{\partial\varepsilon}m_{0k}F - \frac{1}{2}\frac{\partial\alpha}{\partial\varepsilon}p_{0k}F^2 + \frac{1}{2}\frac{\partial^2\alpha}{\partial\varepsilon^2}(m_{0k}F)^2 + \cdots \tag{9.3}$$

In isotropic samples the field vector orientation does not have meaning; the dipole moment and polarizability terms should be understood as orientation averages over all possible orientations, and the linear term in F averages to zero. The first nonzero term is thus quadratic in the field, and the signal is collected at twice the modulation frequency.

From Eq. (9.3) one sees that changes in polarizability are associated with the first derivative of the absorption spectrum (quadratic Stark shift), while changes of the permanent dipole moment are associated with second-order derivatives. Because CT states have large permanent dipole moments, the second derivative contribution to the electroabsorption can be assigned to CT states (second-order linear Stark shift).

The excited state k can undergo field-assisted ionization (or its ionization rate be enhanced in the presence of the field). This adds a decay path for the k-state, shortening its lifetime, and thus leading to line broadening. However, polarization dephasing is in most circumstances much faster than state lifetime, and this effect is too small to be detected. Ionization may appear in time-resolved experiments such as field-assisted pump probe, where the ionization effect gives rise to the buildup of an additional charge population in time.

If the transition 0–n is dipole forbidden, the electric field perturbation may activate it by introducing a privileged direction that breaks the symmetry. The difference spectrum will thus contain a new feature associated with such transition. This qualitative explanation is better understood within quantum mechanics. In quantum mechanics, the effect of the field on the electronic states is described by

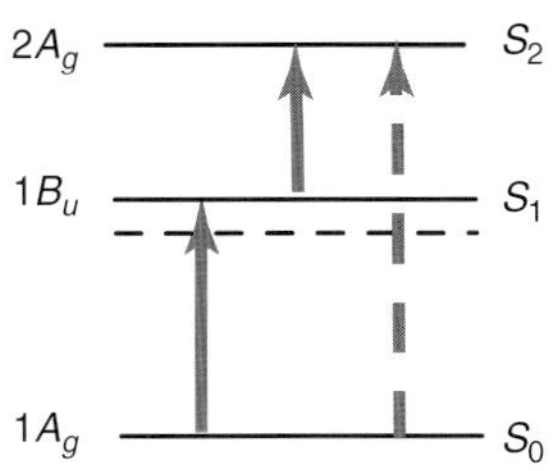

Figure 9.2 Three states model, S_0, S_1, S_2, with wavefunction symmetry A_g, B_u, and A_g, respectively. Accordingly, the only two allowed transitions are $S_0{-}S_1$, and $S_1{-}S_2$. Dashed lines depict the effect of an external electric field: the shift in energy of the $1B_u$ state and the activation of the $2A_g$ state.

perturbation theory. The field mixes the wavefunctions, causing oscillator strength borrowing, and shifts, the energy levels to the second order, according to

$$\psi_k = \psi_k^0 + \sum_{j\neq k} \frac{\left\langle \psi_j^0 \right| \overline{\mu}_{jk} \cdot \overline{F} \left| \psi_k^0 \right\rangle}{E_j^0 - E_k^0} \psi_j^0 \tag{9.4}$$

$$\Delta E_k = \sum_{j\neq k} \frac{\left|\left\langle \psi_j^0 \right| \overline{\mu}_{jk} \cdot \overline{F} \left| \psi_k^0 \right\rangle\right|^2}{E_j^0 - E_k^0} \psi_j^0 = p_k \frac{F^2}{2} \tag{9.5}$$

In Figure 9.2 we consider a simple three-state model, S_0, S_1, and S_2, with wavefunction symmetry A_g, B_u, and A_g, respectively. Accordingly, the only two allowed transitions are $S_0{-}S_1$, and $S_1{-}S_2$. The S_2 state is dipole forbidden from the ground state and will not appear in the linear absorption spectrum. According to Eq. (9.4) the quantum mechanical expression for the perturbed wavefunction in $2A_g$ is $\psi'_{2Ag} = \psi^0_{2Ag} + \beta_{gu}\psi^0_{1Bu}$. This accounts for oscillator strength borrowing because the transition dipole moment $1A_g$-$2A_g$ is now different from zero according to $\mu_{1Ag-2Ag} = \left\langle \psi'_{2Ag} \right| \mu \left| \psi^0_{1Ag} \right\rangle = \beta_{gu} \left\langle \psi^{0'}_{1Bu} \right| \mu \left| \psi^0_{1Ag} \right\rangle$, and the oscillator strength transferred to the forbidden transition is

$$\frac{\Delta f}{f} = \frac{\left|\left\langle \psi^0_{1Bu} \right| \overline{\mu} \cdot \overline{F} \left| \psi^0_{1Ag} \right\rangle\right|^2}{\left(E^0_{1Bu} - E^0_{2Ag}\right)^2} = \frac{\Delta E}{E^0_{1Bu} - E^0_{2Ag}} \tag{9.6}$$

In the last equation, ΔE is the energy shift of the $1B_u$ state due to the field, $\Delta E = \frac{\left|\left\langle \psi^0_{1Bu} \right| \overline{\mu} \cdot \overline{F} \left| \psi^0_{1Ag} \right\rangle\right|^2}{\left(E^0_{1Bu} - E^0_{2Ag}\right)}$, which allows a link with accessible experimental quantities.

The quantum mechanical expression ΔE also accounts for the redshift of the absorption gap, caused by the negative denominator. A nonemitting material with reverse $A_g{-}B_u$ ordering would be consistent with a blueshift of the optical gap.

In general, polarizability describes the response of the molecule to an electric field, and it is strictly linked to spectroscopical and electrical properties. Polarizability is specified by the electronic structure according to the nonresonant formula:

$$p = \frac{2}{3} \sum_n \frac{|\mu_{n0}|^2}{\Delta E_{n0}} \tag{9.7}$$

where there is no frequency dependence, according to the assumption of static field, but transition dipole moments are included. The polarizability volume is defined by $p' = \frac{p}{4\pi\varepsilon_0}$, and it represents the size of the reactive electronic cloud. So a more delocalized state will have a larger volume. A link with spectroscopy can be done by using the oscillator strength in the polarizability definition, $p = \frac{\hbar^2 e^2}{m_e} \sum'_n \frac{f_{n0}}{\Delta E_{n0}^2}$, which leads to the approximated expression, after using the sum rule, $p \cong \frac{\hbar^2 e^2 N_V}{m_e \Delta E^2}$. This shows that a molecule rich in valence electrons (N_v) is highly polarizable. The dielectric constant for an ensemble of N weakly interacting molecules is $\varepsilon_r = \frac{1+2pN/3\varepsilon_0}{1-pN/3\varepsilon_0}$.

Using again the three-level model, the ground-state polarizability, as could be measured by capacitance, is $p(1A_g) = \frac{2}{3}\frac{|\mu_{10}|^2}{\Delta E_{10}}$. Polarizability of the first excited state would be $p(1B_u) = \frac{2}{3}\left(\frac{|\mu_{12}|^2}{\Delta E_{12}} + \frac{|\mu_{10}|^2}{\Delta E_{10}}\right)$. From the electroabsorption experiment the difference $\Delta p = p(1B_u) - p(1A_g) = \frac{2}{3}\frac{|\mu_{12}|^2}{\Delta E_{12}}$ would be accessible.

Finally, because the modulation of absorption due to a static field is a particular situation for the general nonlinear response of the molecule in the presence of an electric field, electroabsorption can also be described by nonlinear susceptibility, $\Delta\alpha = -C \cdot \omega\, \mathrm{Im}(\chi^{(3)})$where, for the three-level model we are considering, the third-order susceptibility is

$$\chi^{(3)}(-\omega; \omega, 0, 0) = \frac{|\mu_{21}|^2\, |\mu_{10}|^2}{(\omega_1 - \omega - i\gamma)^2\, (\omega_2 - \omega - i\gamma)^2} \tag{9.8}$$

Before any advanced theory is applied, a simple fitting of the electro absorption spectrum can be based on the following function:

$$\Delta\alpha(\omega) = A \cdot \alpha + B\frac{\partial\alpha}{\partial\varepsilon} + C\frac{\partial^2\alpha}{\partial\varepsilon^2} + D_n \cdot g(\omega - \omega_n) \tag{9.9}$$

where the first term takes into account the oscillator strength conservation rule and describes a reduction of ground-state absorption (bleaching), the second term accounts for quadratic Stark shift, the third term accounts for CT transitions, and the fourth term describes new absorption bands activated by the field because of symmetry breaking.

Electroabsorption requires the use of electrodes to apply the voltage onto the sample, and it is carried out on test devices. Usually the voltage and electrode configuration is such as to avoid current injection (reverse bias). In sandwich geometry one of the electrodes is semitransparent or transparent to allow optical access. In the presence of different electrodes work function there will be a built-in field in the device. In this situation the modulated applied voltage at ω will display a quadratic Stark effect even when the lock-in detection is at ω. Assume E_0 is the built-in field, and $E_1 \sin(\omega t)$ the applied field. The total field in the sample is $[E_0 + E_1 \sin(\omega t)]$, and its square contains a term $\sim E_0 E_1 \sin(\omega t)$ oscillating at ω. So a Stark effect at first harmonic of the modulation may indicate the presence of a built-in field. Another experimental configuration for electroabsorption is based on interdigited electrodes, which rest on the substrate on top of which the sample is deposited. Note that in the two configurations the field direction with respect to the sample material is different.

In electrofluorescence the applied electric field modulates light emission. The recorded signal is typically the quenching of fluorescence due to ionization of the emitting state by the field. Because emission comes from a well-defined state (the lowest bright singlet), this technique is a specific tool for that state. Usually fluorescence quenching is measured with static field and reported as an overall reduction in emission.

9.2 Pump Probe

A pump probe experiment is the most basic and general approach to time resolution. A stimulus causes a nonequilibrium state, a probe measures this state as it evolves in time (Figure 9.3). This scheme may be used with a variety of excitation and probing phenomena. Here the stimulus and the probe are light pulses; the phenomena involved are electronic or vibrational transitions. A short resonance pulse excites the samples, and a weaker, delayed pulse is used for probing pump-induced changes. Transmission changes are measured on changing pump probe delay, which provides a time-dependent characterization.

There are three levels of knowledge one needs to have for fully controlling and understanding a pump probe experiment:

1) Practical
2) Technical
3) Physical.

Practical knowledge concerns the know-how on the laser source, the optical alignment, and the detection electronics.

Technical knowledge regards radiation–matter interaction, but it does not concern sample photophysics. It can be the simple Lambert–Beer relationship, the χ^3 formalism or the optical Bloch equations. At which level those concepts have to be known depends on the kind of experiment one is doing. Very short pulses, when pulse duration matches electronic dephasing, require a full Bloch equation approach to be described. For longer pulses it is definitively not needed, and for two-color experiments and long pulses the Lambert–Beer formalism is enough.

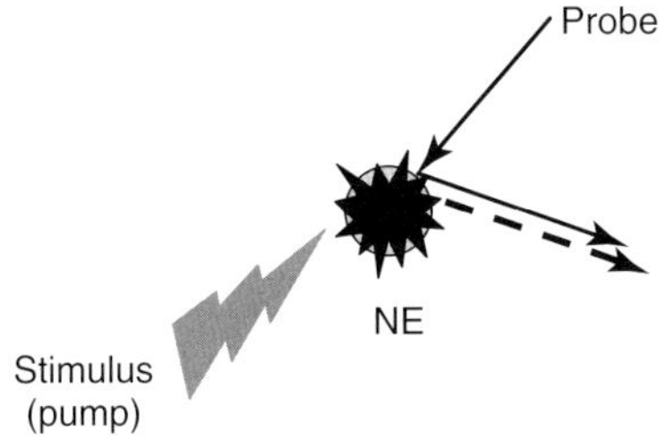

Figure 9.3 The idea of a pump probe. The pump generates a nonequilibrium state. The probe measures its evolution at different times.

The theory for nonlinear optics has been essentially developed in the 1930s and refined, thanks to experiment, in the 1960s and 1970s. All is established, yet the matter is quite complex, and it is not easy to master all its aspects. The same theory explains many pulse generation processes and laser principles.

Physical knowledge regards the photophysics of the material under study, and it is harder to obtain. Pump probe is an advanced experiment that requires full understanding of the linear optical properties, in particular, absorption and emission spectra. It requires a model for the electronic structure, the elementary excitations and their dynamics. On the basis of the results one can build an interpretative model that consists of the photoexcitation scenario for the investigated sample. Sometimes a proper picture is obtained only after years of investigation and the contribution of many research groups. In addition, at variance with point 2, modeling often requires new knowledge, not yet established.

Practical knowledge is by and large acquired in the lab. Some tips and general procedures can be briefly summarized, but direct experience is the only way to learn. The laser source generates a train of pulses, and the number of pulses per unit time is called *repetition rate* (rr), usually expressed in Hertz. The time separation between pulses, which is 1/rr, is the longest delay one can investigate. Longer (slower) phenomena induce pile-up of the signal from one pulse to the other and contribute to a background in the time trace, appearing as a background. Pump and probe are usually obtained from the same laser beam after a number of nonlinear optical operations (harmonic generation, supercontinuum generation, parametric amplification, and Raman shifting). The two trains of pulses, for the pump and the probe, are shifted in time, one with respect to the other, by a delay line (simply a different path in space), of no more than the pulse separation (usually by much less), as in Figure 9.4. This generates a multiplication of one fundamental event: the pump pulse excites the material, and the probe pulse interrogates the pump-induced state at some later time. The observable is the pump-induced change on the probe transmission, reflection, scattering, or polarization state. It is assumed that between two pump pulses the system fully recovers to ground state. This event is reproduced rr times per second and the outcome (e.g., a change in transmission, ΔT) is averaged out. The signal is the normalized change, $\Delta T/T$, versus pump probe delay and probe wavelength. The pump probe delay is fixed during each data point acquisition. The detector time-integrates the transmitted energy, on a very slow time scale, that is many orders of magnitude longer than the pulse width. This is the greatest advantage of pump probe. Time resolution occurs by time correlation between pulses, not by direct time-resolved detection. The limiting factor in time is the pulse time duration. Frequency information can be retraced on a very narrow spectral range, limited by the observation time that is typically in the microsecond time range.

The higher the rr the better the statistical averaging and the signal to noise ratio. However, a high rr has drawbacks:

1) larger chances of long time signal accumulation and thermal effect;
2) average pulse energy is small, and any optical manipulation for wavelength tuning becomes difficult.

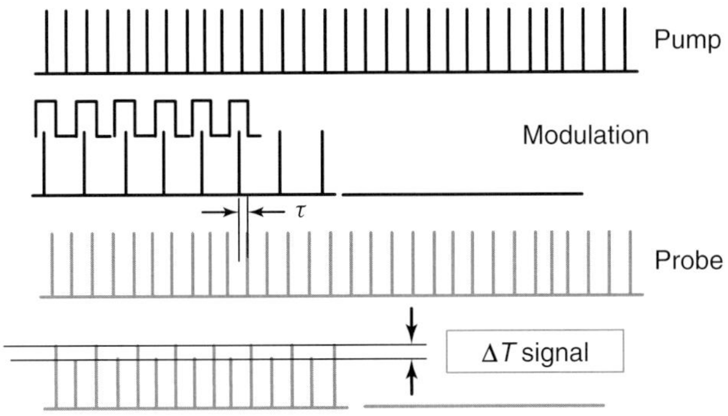

Figure 9.4 A laser beam consisting of rr pulses per second is mechanically modulated at half the repetition rate (for example, to explain lock-in detection) to generate the sequence of pump pulses. The probe beam is a replica, not modulated, shifted in time. The signal is a small modulation, locked in phase and synchronous with modulation, over a large background. Lock-in amplifier allows extracting the weak modulation.

Low rr allows many kinds of nonlinear optic studies because the pulse intensity can be high. Thermal effects are not usually an issue at low rr because there is enough time for dissipation between one pulse and the next; however, the signal to noise ratio is the worst. There is an old Roman saying: *In media stat virtus* (the best is in the center). Depending on what should be measured, the best setup should be chosen.

Pump and probe beams are focused on the same spot (space overlap) and pulses synchronized in time (time overlap). The pump spot is larger than the probe spot to assure overlap even when the pump pulses wander a bit in space, for instance, because of imperfect delay stage or divergence due to distance. Even a perfect setup will suffer by pump divergence on increasing the optical path, for long delays, and such effects can be corrected by calibration using a known sample. The probe pulse should in principle be a small, negligible perturbation, and it should have a much smaller intensity than the pump. The pump pulse transmitted through the sample should not reach the detector. Space filtering, spectral filtering, and polarization analyzer are known strategies to avoid this. However, it cannot be fully avoided. The pump pulse scattered light does not depend on the pump probe delay and contributes a constant background in time traces. Lock-in amplification is one of the fundamental approaches for detection. The pump beam is modulated, and changes in the probe transmission are selected at proper frequency and phase with respect to modulation. A difference-detection system is sometimes adopted. Two probe beams are collected by two equal detectors and their signal difference tuned to zero. One of the probe pulses is aligned to go through the excited spot of the pump, thus carrying the signal. Changes in the photodiode balance are detected as pump-induced effects, while other laser fluctuations, equal on both probe beams, are nullified by the difference. Whole spectra can be acquired by registering pump-on and pump-off probe spectra onto diode arrays or CCDs by using a broadband probe (for instance, supercontinuum). Proper subtraction

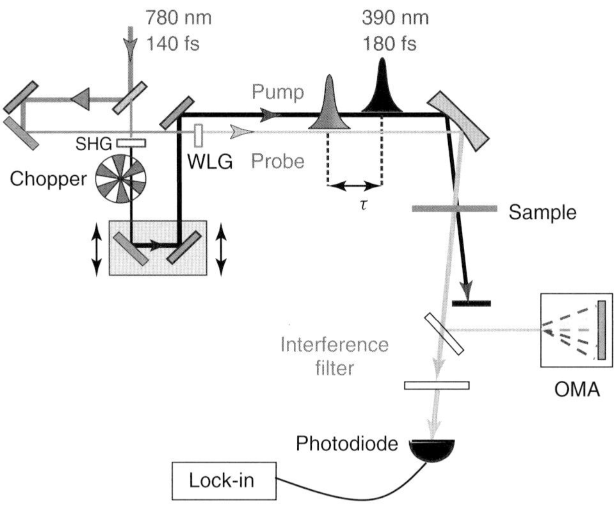

Figure 9.5 A schematic layout of a pump probe setup. SHG, second harmonic generation; WLG, white light generation; OMA, optical multichannel analyzer.

techniques provide the signal as normalized transmission. An example of a pump probe experimental layout is given in Figure 9.5.

Technical knowledge relies on the well-established theory for radiation–matter interaction, and it can be stated at various degrees of approximation. Photoexcitation changes the optical properties of a sample. Let us focus our attention on transmission, T. In presence of pump and probe fields, there is a total nonlinear polarization in the material, represented by the expansion $P(t) = P^{(1)}(t) + P^{(3)}(t) + \cdots$ Here we are concerned with the nonlinear part of the polarization, $P^{(3)}(t)$. There are several ways to detect a polarization.

1) By measuring the normalized loss in energy of the transmitted probe due to beating with the nonlinear polarization, $S(\tau) = \frac{\int E(t-\tau)\cdot \frac{\partial P^{(3)}}{\partial t}\,\mathrm{d}t}{\int |E(t)|^2 \cdot \mathrm{d}t}$. This is the usual loss mechanism, as seen with linear response in Chapter 2.The signal depends on the *imaginary part* of the nonlinear susceptibility. In the experiment the probe pulse is integrated in time and spectrum.
2) By measuring the Fourier transform of the loss term, $S(\tau, \omega) = \frac{\omega L}{\varepsilon_0 nc}\,\mathrm{Im}\,\frac{P^{(3)}(\omega)}{E'_{\mathrm{p}}(\omega)}$. Here the probe pulse is dispersed and spectrally resolved.
3) By measuring the emitted power $H(\tau) \propto \left|P^{(3)}\right|^2$, either time integrated or time resolved, collected at specific direction according to wave vector conservation.

In a pump probe transmission experiment the collected signal is the change in energy of the transmitted probe. In order to understand why we should consider that the nonlinear polarization leads to the generation of an additional field component in the probe direction. Within the slowly varying amplitude approximation for the wave equation (see Appendices 9.A and 9.B) this field is

$E_R(t) = -\frac{L}{2n\varepsilon_0 c}\frac{\partial P^{(3)}}{\partial t}$ for a sample thickness L, refractive index n. The transmitted probe intensity is $I(\omega) = \left|E'_p(\omega) + E_R(\omega)\right|^2 \approx \left|E'_p(\omega)\right|^2 + 2\,\mathrm{Re}\,E'^*_p(\omega)E_R(\omega)$ where E_R is assumed small and E'_p is the probe field modified by the linear polarization. Note that the last term is of the kind $\propto E^*_p\frac{\partial P^{(3)}}{\partial t}$, that is, the energy lost by the probe field because of the nonlinear polarization (see Chapter 2). The normalized change in transmission, at fixed pump probe delay, is

$$\frac{\Delta I(\omega)}{I_0(\omega)} = \frac{\Delta T}{T}(\omega) \approx \frac{2\,\mathrm{Re}\,E'^*_p(\omega)E_R(\omega)}{\left|E'_p(\omega)\right|^2} = \frac{\omega L}{\varepsilon_0 nc}\,\mathrm{Im}\,\frac{P^{(3)}(\omega)}{E'_p(\omega)}$$

Because $P^{(3)}(\chi^3)$ depends on three fields, many "nonlinear" processes can contribute to the signal, which are not the intuitive and fundamental population effect one has in mind. Among such phenomena there is a four-photon parametric conversion, stimulated Raman scattering, self phase modulation, self-steepening, coherent coupling, pump-perturbed free induction decay, two-photon absorption, and more. Most of these processes, however, take place during pulse overlap in time. In the well-separated pulse situation, only one term survives. This is called *population term*, and it can be understood without using the third-order susceptibility formalism, even if it is a manifestation of that. First, we discuss this regime, where the nonlinear response function mentioned above is taken into account by an effective linear formulation of the transient transmission, embodied in time-dependent population.

According to the Lambert–Beer relationship, transmission T is given by

$$T = \mathrm{e}^{-\alpha L} \tag{9.10}$$

where α is the absorption coefficient and L is the sample thickness. Absorbance A is defined as

$$A = -\log_{10} T \tag{9.11}$$

The frequency-dependent absorption coefficient is given by a sum of all possible transitions between pairs of states, $i.j$, with population N_i and N_j according to

$$\alpha = \sum_{i,j}\sigma_{ij}(\omega)(N_i - N_j) = \sum_j\left(\sum_i\tilde{\sigma}_{ij}(\omega)N_j\right) \tag{9.12}$$

where σ_{ij} is the cross section of the transition. This quantity is defined positive, but one can write a compact expression with an effective cross section, which takes a negative sign for downward transitions and a positive sign for upward transitions, so that only the starting population is explicit. This is a valid practical simplification that can be done when final states of the transitions are empty.

Now the pump pulse is introduced, at time $t = 0$, causing the transmission change from T to T^*. Accordingly,

$$\Delta A = -\log\left(\frac{T^*}{T}\right) = -\log\left(1 + \frac{\Delta T}{T}\right) \cong -\frac{\Delta T}{T2.3} \tag{9.13}$$

where $\Delta T = T^* - T$ and the last approximation holds for small signals.

Now let us assume that the population of the k-state is changing because of pump excitation according to

$$N_k^* = N_k + \Delta N_k(t) \tag{9.14}$$

The probe pulse interrogates the nonequilibrium state at time τ. The measured change in absorbance is

$$\Delta A(\tau) = -\log\left\{\frac{\int dt I_{\mathrm{pr}}(\omega, t-\tau)\exp\left[-\sum_{kj}\tilde{\sigma}_{kj}(\omega)\Delta N_k(t)L\right]}{\int dt I_{\mathrm{pr}}(\omega, t)}\right\} \tag{9.15}$$

which, for a small signal, can be approximated to

$$\frac{\Delta T}{T} = -\frac{\int dt I_{\mathrm{pr}}(t-\tau)\left[\sum_{kj}\tilde{\sigma}_{kj}(\omega)\Delta N_k(t)L\right]}{\int dt I_{\mathrm{pr}}(t)} \tag{9.16}$$

If we assume that the probe pulse is much shorter than the characteristic time scale of evolution of ΔN, that is, the probe pulse is a δ-like pulse, then the expression is even simpler:

$$\frac{\Delta T}{T}(\omega, t) = -\sum_{i,j}\tilde{\sigma}_{ij}(\omega)\Delta N_j(t)L \tag{9.17}$$

This equation rests at the heart of the pump probe experiment: the resonant pump causes population redistribution among some of the electronic states of the sample, creating a nonequilibrium state. The probe measures new transitions associated with such population changes and tracks in time the recovery of the system to equilibrium.

If we resume the positive cross section as in Eq. (9.3), we get

$$\frac{\Delta T}{T}(\omega, \tau) \cong -\sum_{i,j}\sigma_{ij}(\omega)\left[\Delta N_i(I_{\mathrm{pu}}, \tau) - \Delta N_j(I_{\mathrm{pu}}, \tau)\right]L \tag{9.18}$$

When the probe pulse duration cannot be disregarded, Eq. (9.17) or (9.18) should include the correlation with the pulse temporal profile (in intensity) f_{P}:

$$\Delta T/T(\omega, \tau) = -L\sum_{i,j}\tilde{\sigma}_{ij}(\omega)\Delta N_J(\tau)\otimes f_{\mathrm{p}} \tag{9.19}$$

Note that the time-dependent population term includes the pump pulse behavior, which is in many cases represented by a convolution between the material response and the pump pulse profile.

By using pulses of duration t_{P} processes occurring on time scales $\Delta t > t_{\mathrm{P}}$ can be reasonably studied. With noise-free measurements (but they do not exist) you could in principle track down processes occurring on shorter time scales, by carefully deconvoluting the time traces once both the pump and probe temporal profiles are known with very high accuracy. This does not happen in reality, and the suggestion is as long as possible to study dynamics occurring on time scales longer than the pulses available in your experiment.

To understand the role of the pulse duration on the experimental resolution, we can use a simple expression based on pulse intensities.

Let $\Delta\alpha$ be the change in absorption (or any other property we like to measure) induced by the pump pulse. Then we have

$$\Delta\alpha(t) = A(t) \otimes I_{\mathrm{pu}}(t) = \int_{-\infty}^{t} A(t-t')I_{\mathrm{pu}}(t')\mathrm{d}t' = \tilde{A}(t)\tilde{I}_{\mathrm{pu}}(t) \tag{9.20}$$

where $A(t)$ is the response function we like to measure. The last term is the product of the Fourier transform functions that correspond to convolution of the original functions.

If the probe pulse has a finite time duration, then we will measure the correlation of the above signal with the probe pulse intensity temporal profile, $S(\tau) = \Delta\alpha(t) \oplus I_{\mathrm{pr}}(t-\tau) = \Delta\alpha(t)\tilde{I}^{*}_{\mathrm{pr}}(t)$, where again we use the property of the Fourier transform. The overall signal, after simple manipulation, will be

$$S(t) = A(t) \otimes G(t) = \int_{-\infty}^{t} A(t-t')G(t')\mathrm{d}t' \tag{9.21}$$

where $G(t) = I_{\mathrm{pu}}(t) \oplus I_{\mathrm{pr}}(t) = \int_{-\infty}^{\infty} I_{\mathrm{pu}}(t)I^{*}_{\mathrm{pr}}(t-\tau)\mathrm{d}t'$ is the cross-correlation of the pulses.

Although simplified, because the true response function is expressed on fields and not on intensities, this expression is of help. It clearly shows that time resolution depends on the cross-correlation of pump and probe pulses.

Let us now discuss how we can describe the material evolution under photoexcitation. The pump pulse acts on the sample by changing the stationary level occupation, $N_i \rightarrow N_i(t)$ where t is the time and N_i the equilibrium population (which is usually zero, except in ground state). With appropriate initial conditions, and neglecting the coherent regime that would require Bloch equations, the time-dependent population can be obtained from rate equations such as

$$\frac{\mathrm{d}N_i(t)}{\mathrm{d}t} = G_i(t) - R_i(t) \tag{9.22}$$

where $G(t)$ and $R(t)$ are the generation and deactivation rates, respectively. For instance, if the state "i" is directly populated by the pump pulse via one-photon transition from the ground state, the generation term for a thin sample is

$$G_i(t) = \sigma_{0i}N_0(t)\frac{F(t)}{t_{\mathrm{P}}} \tag{9.23}$$

where σ_{0i} (cm^2) is the cross section for the one-photon transition $0 \rightarrow i$; $N_0(t)(\mathrm{cm}^{-3})$ is the time-dependent ground-state population, which gets depleted; $F(t)$ ($\mathrm{photon{\cdot}cm}^{-2}$) is the pump photon flux; and t_{p} (s) is the pump pulse duration.

Solving one rate equation for each involved state "i" allows to reconstruct the signal according to Eq. (9.18) where $\Delta N_i(t) = N_i(t) - N_i$.

The spectrum associated with the state (or species) "j" is $A_j(\omega) = \sum_i \sigma_{ij}(\omega)d$, comprising all possible transitions from this state, and Eq. (9.8) can also be written as $\frac{\Delta T}{T} = -L\sum_j A_j(\omega)N_j(t)$, an expression often used in global fitting analysis. In this analysis one: (i) gets population kinetics from rate equations, (ii) associates with each population a spectrum, and (iii) reconstructs the expected signal. This formula shows that the typical $\Delta T/T$ spectrum is a superposition of individual spectra ($A_j(\omega)$) from several photoexcited states, and time-dependent data must be taken at various probe wavelengths in order to single out the various contributions. Note that global fitting may be be carried out by "blind" routine, which finds mathematically the best combination of spectra. Such results mostly lack physical meaning and should be considered with great caution. The "guess" on the excited species in the experiment should be assessed by general knowledge on the system and cross-checked with other experiments or by varying the experimental condition (see Box 9.1).

Box 9.1: About Fitting and Interpretation

William of Ockham was an English Franciscan friar and scholastic philosopher who lived in the Middle Ages (c. 1288–c. 1348). Possible Umberto Eco was inspired by him in writing "Il nome della rosa." One important contribution that he made to modern science and modern intellectual culture was the principle of parsimony in theory that came to be known as Ockham's Razor. In one of the many enunciates the principle says: "Non sunt multiplicanda entia sine necessitate et frustra fit per plura quot potest fieri per pauciora" (one should not multiply entities beyond necessity, and it is useless doing with the many what can be done with the few). Beside academic discussions about if he ever said that in those words or what far and deep consequences this may have in ontology or theology or philosophy, here we do a brave jump and re-state the principle as *One should not introduced fitting parameters without need and keep the model as simple as possible.* Probably it is not always true that the simplest explanation is also the correct one, yet the principle of parsimony has a great value in physics. At least, one should avoid introducing hypothesis that he cannot support with solid evidence, even if those hypothesis he likes, would make him famous and would improve fitting of experimental data.

Fitting in materials science is an art, with some connection to statistics. However, mathematics is not at the heart of it. Because of this, the danger of introducing too many parameters exists and jeopardizes any interpretation. The more fitting parameters one has, the better will be the fitting, the less it will be the physical meaning.

The principle to follow could be one phenomenon, one experiment. So if you think three phenomena are involved, let's find three independent experiments, each pointing to one of those phenomena, separately or together. This provides the solid evidence I mentioned above.

The worst thing you can do is running a fitting procedure without having a clear physical picture in mind. That will drive you all the way out of reality, in spite of a perfect matching between data and simulation. To sum up, numerical analysis is a good servant, but it can be a bad master.

This box is dedicated to my esteemed colleague and friend Larry.

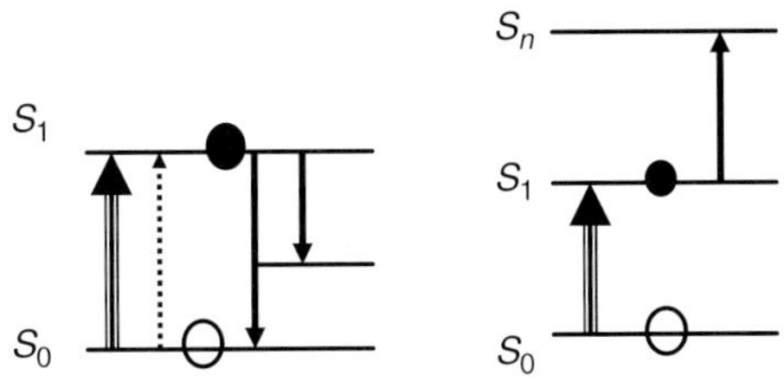

Figure 9.6 Illustration of photobleaching (a), stimulated emission (a), and photoinduced absorption (b). Solid circles are occupied states. Open circles are empty states. The double-line arrows represent pump-induced transitions. The simple arrows represent probe-induced transitions. The dashed line arrow represents a weaker intensity transition, due to population depletion.

For $j = 0$, $\Delta N_0(t)$ describes ground-state depletion and corresponding enhancement in probe transmission, named photobleaching (PB). PB corresponds to positive ΔT and has the spectral shape of ground-state absorption when thermalization, which is usually very fast, is over. For all the other levels ($j > 0$), the probe pulse can stimulate both upward and downward transitions. In particular, transitions to higher lying states give rise to photoinduced absorption (PA) and transitions to lower lying states give rise to stimulated emission (SE) (Figure 9.6). After thermalization, usually very fast, few lower lying states are occupied. When the lowest singlet excitation (S_1) is dipole-coupled to S_0, that is, the system is luminescent, SE takes place. In this case, the relationship between the SE spectrum and the photoluminescence (PL) spectrum can be derived from A and B Einstein coefficients as $\mathrm{SE}(\nu) = \frac{c^3}{8\pi h}\frac{1}{\nu^3}\mathrm{PL}(\nu)$. In general, the existence of SE does not mean that positive ΔT is observed in the region of emission, for it depends on the spectral overlap with other absorbing transitions (often there) and on the relative cross sections involved. As a result, SE may not at all appear, even in light-emitting materials. Another important observation is that SE is clearly distinguished when the absorption edge is sharp and the transition occurs from the excited state to the vibrational replica of the ground state. When SE occurs to the vibrationless ground state, this fully overlaps with absorption and "contributes" to PB. It becomes difficult to distinguish if enhanced transmission is indeed due to light amplification or just reduced absorption. Both are positive ΔT signals, but light amplification (gain) requires population inversion. To check this, one can measure the linear ground-state absorption, α, and compare it to the nonlinear pump-induced change in absorption $\Delta\alpha = -\frac{1}{L}\ln\left(1 + \frac{\Delta T}{T}\right)$. If $-\frac{\Delta\alpha}{\alpha} > 1$, there is light amplification and gain.

As an example of the use of rate equations, let us consider a simple dynamics: a molecule is excited to the first singlet state, then by spin flip it undergoes intersystem crossing to the triplet state. The levels are as in Figure 9.7. The rate

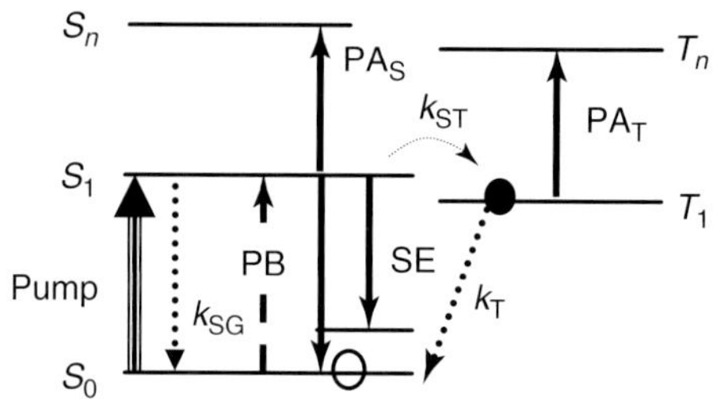

Figure 9.7 The simplified Jablonsky diagram for intersystem crossing on photoexcitation. Rates are indicated by thin dotted lines and optical transition by solid lines. Photobleaching is represented by dashed arrow. The double arrow represents excitation.

equations are

$$\frac{dN_G(t)}{dt} = -G_i(t) + k_{SG}N_S(t) + k_T N_T(t)$$

$$\frac{dN_S(t)}{dt} = G_i(t) - k_S N_1(t)$$

$$\frac{dN_T(t)}{dt} = k_{ST}N_S(t) - k_T N_T(t) \tag{9.24}$$

where N_G, N_S, and N_T are the populations of the ground state, first singlet state, and triplet state, respectively. $K_S = K_{SG} + K_{ST}$ is the total decay rate from S_1, K_{ST} is the intersystem crossing rate, and K_T the deactivation rate of the triplet back to the ground state. The initial condition for the three populations is $[N_G(0), N_S = 0, N_T = 0]$.

The pumping rate is described by Eq. (9.23), where $F(t)$ is the pulse temporal envelope. When the time constants of the all the processes are much longer than t_p, this term can be neglected, changing the initial condition into $[N_G(0) - N_{ex}, N_{ex}, 0]$. N_{ex} is the concentration of absorbed pump photons $N_{ex} = (1 - R_p)\left(1 - e^{-\sigma_{01} N_0 L}\right) F_p/L$, where R is the reflectivity at pump wavelength. When all pump photons are absorbed (thick sample) $N_{ex} \approx \sigma_{01} N_0 F_p$.

The deactivation rate of the triplet usually occurs on the nanosecond to microsecond time scale and can be disregarded when the probed temporal range is much shorter. According to the rate Equation (9.24), the triplet population builds up exponentially at time rate K_S, that is, the triplet population forms on the same time scale of singlet decay. The yield of triplets is given by the efficiency of the process, $\eta_{ST} = k_{ST}/k_S$. In carbon-based conjugated molecules k_{ST} is in the order of ns^{-1}, thus what matters for getting a sizable triplet population is the decay rate of the singlet, k_S. If the only deactivation channel is singlet to triplet conversion, $k_S = k_{ST}$, then $\eta_{ST} = 1$. In general, $\eta_{ST} < 1$. This explains why spin flip is not effective for higher lying singlet states. The latter have lifetime in the order of 100 fs, and $\eta_{ST} = 10^{-4}$. This tiny triplet population forms anyway in the time scale of decay of the singlet, in this last situation, 100 fs. So spin flip can take place in a very short time, but usually with very small yield. More on this can be found in the discussion of the Jablonsky diagram.

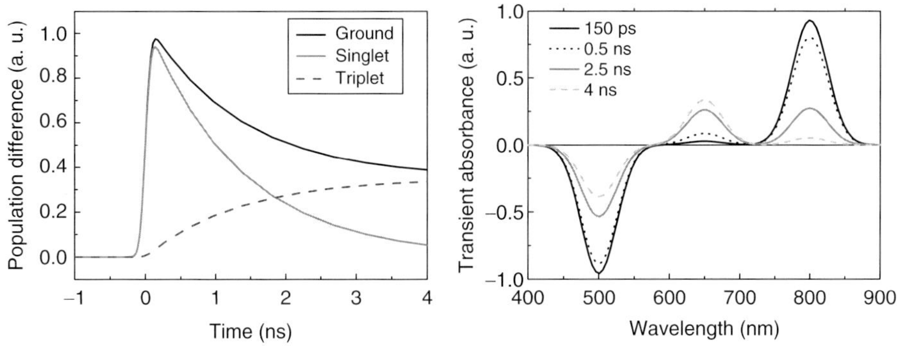

Figure 9.8 Numerical simulation based on Eqs. (9.15) and (9.16). On the left we plot population kinetics. "Ground" regards the change in ground state population ($N_G(t) - N_G$) and its recovery. Singlet and triplet are the respective state population, $N_S(t)$ and $N_T(t)$. On the right we plot "transient absorbance" $\left(-\frac{\Delta T}{2.3T}\right)$ spectra that are reconstructed using gaussian lineshape. All transitions have same cross-section. The negative peak at 500 nm is PB, while the two positive bands at 650 and 800 nm are the triplet–triplet and singlet–singlet absorption, respectively.

Once population kinetics are known, the pump probe signal has to be reconstructed using

$$\Delta A = -\frac{\Delta T}{2.3T}(\omega, t) = -L\sum_j A_j(\omega)N_j(t)$$
$$= -L\left\{A_G(\omega)(N_G - N_G(t)) - A_S(\omega)N_S(t) - A_T(\omega)N_T(t)\right\} \qquad (9.25)$$

where the first terms accounts for PB, the second for singlet–singlet absorption, and the third for triplet–triplet absorption. Figure 9.8 shows a numerical solution of Eqs. (9.24) and (9.25) using a Gaussian pump pulse of 150 ps and three identical Gaussian lineshape bands for ground-state absorption, singlet–singlet absorption, and triplet–triplet absorption. The singlet internal conversion (IC) rate to ground state is $K_{SG} = \frac{1}{2}\ \text{ns}^{-1}$, the intersystem crossing rate is $K_{ST} = \frac{1}{4}\ \text{ns}^{-1}$, and the triplet lifetime K_T is assumed infinite. As a consequence, the total deactivation rate of the singlet is $\frac{3}{4}\ \text{ns}^{-1}$ and the triplet yield is $\eta_{ST} = k_T/k_S = 1/3$.

To conclude, we note that the example has a number of rough approximations. Absorption lineshapes are in reality the convolution of a distribution function (inhomogeneous broadening) with vibronic manifolds. The deactivation rates are very often dispersive, that is, time dependent, because of a distribution of lifetimes in the inhomogeneous band, so that decay is rarely pure exponential.

The pump probe technique can be used for monitoring excited-state migration in space by measuring the photoinduced anisotropy loss in time. This is possible when the linearly polarized pump induces optical anisotropy in an otherwise isotropic sample. For instance, if the optical transition in sample molecules is linearly polarized, the initial orientation of the excited molecules will be preferentially along the pump field. More precisely the molecular orientation distribution will follow a square cosine law on the with angle between the molecular transition

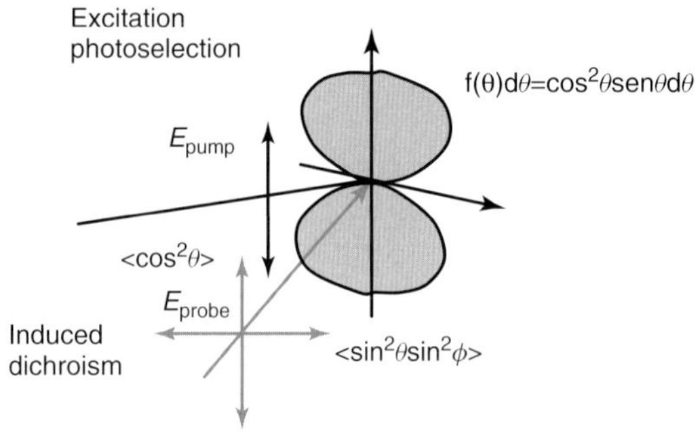

Figure 9.9 The gray area describes the initial distribution in space of the angle between the excited molecule transition dipole and the pump field. This distribution follows the cosine square law. A probe field parallel to the pump field will suffer an interaction proportional to the fourth power of the cosine, while the perpendicular probe will see a square sine per square cosine product. This determines a difference in the two signals.

moment and the pump field. In this situation the probe pulse, polarized parallel or perpendicular to the pump field, allows keeping track of the induced anisotropy, ρ, defined as $\rho = \frac{\Delta T_{//} - \Delta T_{\perp}}{\Delta T_{//} + 2\Delta T_{\perp}}$ (Figure 9.9). On exciton motion in a disorder medium, where molecular orientation is not retained, the initial pump-induced orientation is lost ($\rho \rightarrow 0$), providing a direct measure of the time scale of the process. Note that the quantity $\Delta T_{//} + 2\Delta T_{\perp}$is proportional to the total population in the sample, disregarding orientation.

Each pump probe signal, $\Delta T_{//}/T$ and $\Delta T_{\perp}/T$, has a spurious decay, due to a combination of diffusion and population decay. The evaluation of $\rho(t)$ as defined above allows disentangling those two contributions. One can skip the two measurements and the evaluation of anisotropy by performing the pump probe experiment with pump probe polarization at "magic angle," 54.7°. In this configuration, diffusion cannot affect the pump probe signal. In Box 9.2 we briefly mention about a technique using three beams, pump-push-probe. This experiment is formally equivalent to a fifth order non linear scattering, and it can sometime be useful to explore excited state properties.

Box 9.2: The Vavilov–Kasha Rule and the Pump-Push-Probe Experiment

In pump-push-probe, a first pulse excites the sample (pump), a second pulse reexcites the sample, and a third pulse probes the induced changes. The pump pulse is resonant with the material absorption, and the second pulse is tuned at the excited-state transition. The third pulse in the example below probes the singlet state population S_1 by measuring SE (see Figure 9.10). Here the push pulse arrives about 0.5 ps after the pump. The probe scans all delays from negative to positive. Without push, the time trace shows SE. With push, there is a very large dip in SE, due to S_1 depletion caused by reexcitation to S_n. The technique allows investigating the Vavilov–Kasha rule. The recovery from S_n

to S_1 by IC is ultrafast, almost not time resolved (150 fs) and almost complete. A small fraction of the S_n population, however, does not come back to S_1. Other experiments show that this population is lost into the charge generation channel, giving rise to interchain, long-lived polaron pairs.

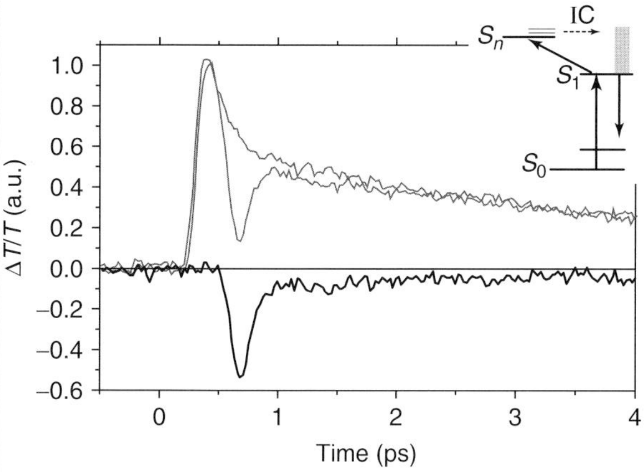

Figure 9.10 Stimulated emission with and without push. The thick negative black trace is obtained from difference of the two above, and it highlight the recovery of S1 population aftef push re-excitation. In the inset the involved transitions. After C. Gadermaier *et al.*, (2002).

9.3 cw Photoinduced Absorption

In this experiment, a cw monochromatic light resonantly excites the sample. The excitation is mechanically modulated at frequencies typically between 10 Hz and 10 kHz. During illumination, the sample reaches a metastable equilibrium, referred to as *quasi steady state*, while in the dark phase it recovers back to the initial equilibrium. A broadband incoherent cw light, usually provided by a lamp, is used to measure the change in transmission caused by the photomodulation. The technique is similar to pump probe, and it measures a transmission difference usually via lock-in detection, normalized as $\Delta T/T$. Figure 9.11 shows a scheme of the setup.

Let us assume that the modulated excitation is described by a square wave of period T_M and frequency $\omega_M = \frac{2\pi}{T_M}$ (rads^{-1}). The population N of a state with lifetime τ will build up during excitation to a value that depends on T_M. This is shown in Figure 9.12 for two examples. When $\tau \ll T_M$ the population will reach a steady state level. By lowering the modulation frequency (increasing T_M) one should be able to reach the maximum signal, corresponding to steady state, and from that point the cw photoinduced absorption (cwPA) will be frequency

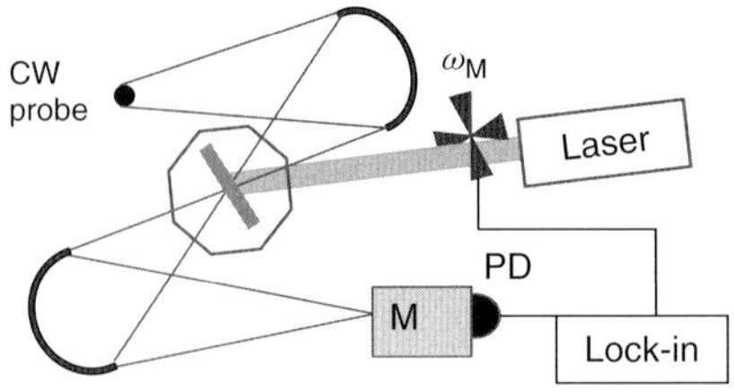

Figure 9.11 cw Photoinduced absorption setup.

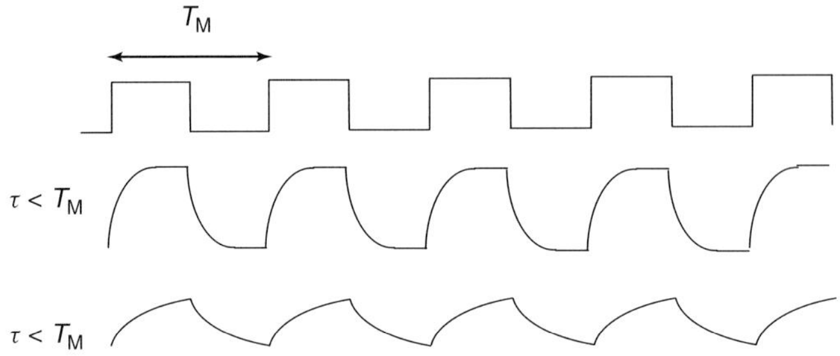

Figure 9.12 (a) Square wave excitation in a cwPA experiment. (b) Signal kinetics for lifetime shorter than modulation period; a steady state is reached at any cycle. (c) Signal kinetics when lifetime is longer than the modulation period. The signal does not reach steady state. The lock-in amplifies the amplitude of the signal component at the first harmonic of the modulation.

independent. When $\tau > T_M$ the photoinduced population will reach only a fraction of that value, that tends to zero for high frequencies. This dynamics is described by a simple differential equation:

$$\frac{dN}{dt} = G(t) - \frac{N}{\tau} \tag{9.26}$$

where $G(t)$ is the generation term and N/τ the monomolecular recombination term. In the rate equation we assume a sinusoidal generation rate $G(t) = G_0(1 + e^{i\omega_M t})$, such that the real part of $G(t)$ oscillates between 0 and 1 (dark and light) and G_0 specifies generation according to $G_0 = \eta(1 - R)(I_0 - I_T)/h\nu$, where η is the quantum generation efficiency, $h\nu$ is the pump photon energy, R is the reflectivity, T is the transmission, and I is the incident (0) and transmitted (T) pump intensity. The solution of Eq. (9.26) is

$$N(t) = e\frac{t}{\tau}\left[C + G_0 \int dt\left(1 + e^{i\omega_M t}\right) e\frac{t}{\tau}\right] \tag{9.27}$$

After integration

$$N(t) = G_0\tau + \frac{G_0(1 + e^{i\omega_M t})}{1/\tau + i\omega_M} + Ce^{-\frac{t}{\tau}} \tag{9.28}$$

Besides a constant and a transient term that decays off, the population $N(t)$ oscillates at frequency ω_M and can be detected using the lock-in technique. The oscillating term is

$$N_\omega(t) = G_0\tau\frac{e^{i\omega_M t}}{1+i\omega_M\tau} = G_0\tau\frac{e^{i(\omega_M t+\phi)}}{1+\omega_M^2\tau^2} = G_0\tau\frac{1-i\omega_M\tau}{1+\omega_M^2\tau^2}e^{i\omega_M t} \quad (9.29)$$

In this expression $tg\phi = \omega_M\tau$ is the phase of the signal with respect to the photo modulation. The lock-in signal is the phase-dependent amplitude of the oscillation, and it will have a real, in-phase component and an imaginary, out-of-phase (quadrature) component. The phase of the signal is proportional to the state lifetime, and a long-lived state will have a predominantly quadrature component, while a short-lived state will be in phase.

Note that the measured signal will also bear an instrumental phase shift, without physical meaning, because of the modulation mechanism (e.g., how the chopper blades cut the light beam). This phase shift is easily evaluated by recording the exciting scattered light and should be subtracted.

For $\tau \ll T_M$ steady state is reached, $\frac{dN}{dt} = 0$ and $\tilde{N} = G_0\tau$.

With mechanical modulation (10 Hz to 10 kHz), the range of accessible lifetime is 0.1–100 ms. In organic semiconductors, such long-lived photoexcitations are typically trapped charge carriers or triplet states. Also in semiconductor nanocrystals, cwPA may be due to charges trapped at the surface. As in pump probe, the signal can be positive (bleaching) or negative (PA), according to $\frac{\Delta T}{T}(\lambda, t) = -\sum_j A_j(\lambda)\tilde{N}_j(\omega_M)L$, where L will be the sample thickness for thin samples, or $1/\alpha_P$ for thick samples, where α_P is the absorption coefficient at pump frequency. Here λ is the probe wavelength that regards the transmission spectrum, and A_j is the multiline cross section associated with the state j. This cross section may contain several transitions at different energies that represent the spectrum of the photoexcitation "j." In films of conjugated polymers, for instance, the cwPA spectrum at a low temperature shows several PA bands assigned to trapped polarons, bipolarons, or triplet states. At very long wavelengths, in the vibrational infrared spectral region there are sometimes sharp features, assigned to IR active vibrations (IRAV) associated with charged excitations. These bands originate from the symmetry breaking occurring at the trapped charge site that activate Raman modes. As a consequence, the IRAV frequencies correspond to the Raman modes of the polymer chain. The trapped charges oscillate under excitation, "donating" nonnull dipole moment to the Raman transition (see Section 5.4.3). In the presence of trapped charges there is another interesting effect, observed in the bleaching spectral region. Trapped charges may induce a strong local electric field. This in turn causes electroabsorption. Owing to this mechanism it is sometime possible to see EA modulation in cwPA spectra.

The dynamics or kinetics of the long-lived photoexcitations can be investigated by changing the experimental conditions. cwPA data can be acquired at different temperatures, excitation intensities, excitation photon energies (excitation profile), and modulation frequencies.

There are two temperature effects we can talk about. One is the different kinetics at different temperatures of the photoexcited species. This is studied by taking cwPA spectra at different temperatures. The other is unavoidable heating of the sample during photoexcitation. Any photoexcitation causes some thermal heating. Usually the increase in temperature is small, yet not necessarily negligible. The increase in temperature upon photoexcitation can be estimated from the absorbed photon energy. Let us assume Q is the energy delivered to the sample in one second per unit area (this can be as well adopted for pulsed excitation to represent the absorbed energy per unit area per pulse). Using the relation $Q = C_P \rho L \Delta \tilde{T}$, where C_P is the thermal capacity of the sample, ρ is the density, and L is the sample thickness, the temperature increase $\Delta \tilde{T}$ can be estimated. Typical order of magnitude for organic semiconductor films are $\rho \approx 1\text{g/cm}^3$ and $C_P \approx 1\text{J/gK}$. Once the increase in temperature has been estimated, one needs to know the thermal modulation coefficient $\frac{\Delta A}{\Delta T} = \frac{1}{T}(\Delta T/\Delta \tilde{T})$, which links the change in temperature to the change in transmission or absorbance. This can be measured by thermal modulation spectroscopy, an experiment in which a proper sample holder (typically a thermoresistor) modulates the temperature of the sample or, more simply, (and roughly) by taking an absorption spectrum at two different temperatures (e.g., room temperature and liquid nitrogen) and working out their difference.

The lifetime of the photoexcited species can be studied by changing ω_M. On changing the modulation frequency the in-phase cwPA signal will show a plateau at low values, a transient region (appearing as a knee in the plot PA vs ω_M), and then a decay. The out-of-phase signal will show a bell-shaped spectrum. In Figure 9.13 we plot the in-phase (real) and quadrature (imaginary) component as obtained from Eq. (9.29). Both the "knee" and the peak, for in-phase and quadrature, respectively, indicate the lifetime of the photoexcited species. All this is true if a single species is excited and there is monomolecular decay. If a distribution of lifetimes is present in the sample, the behavior described above is no longer true, to a degree that depends on how far the experimental conditions depart from the single lifetime approximation.

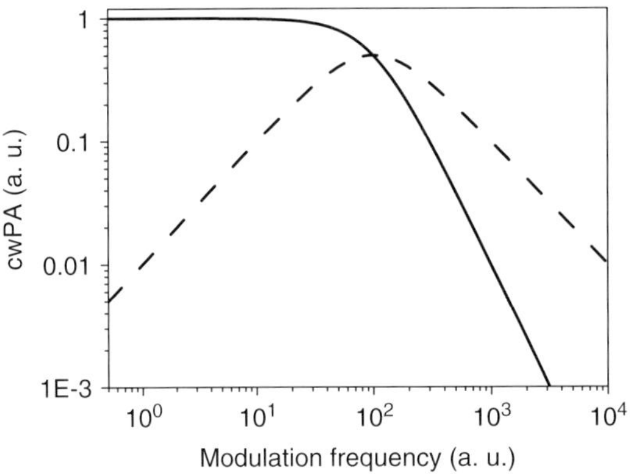

Figure 9.13 Plot of the real in-phase (solid) and imaginary quadrature (dashed) parts of Eq. (9.29). The knee in the real trace and the peak in the quadrature approximately indicate when $\omega_M \tau = 1$, that is, the lifetime of the excited state (there is a 2π in the definition of modulation frequency).

Box 9.3: Trap-Mediated cwPA

Let us assume that long-lived states seen in cwPA come only from trapped states with a finite density N_T. Their number N will grow under limitation, showing saturation on increasing the excitation intensity, I as described in Figure 9.14. The rate equation will be

$$\frac{dN}{dt} = G(N_\tau - N) - \frac{N}{\tau}$$

At steady state this provides $G(N_\tau - N) = \frac{N}{\tau}$ and then

$$N = \frac{G\tau N\tau}{(G\tau + 1)} = \frac{AI}{BI + 1}$$

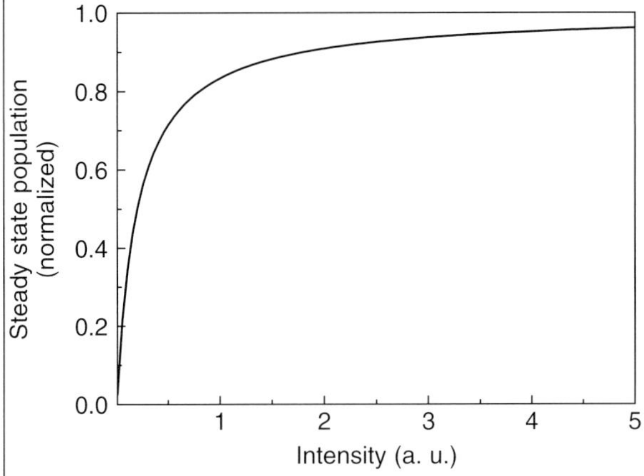

Figure 9.14 The behavior of the cwPA signal upon increasing excitation intensity for a trap-limited regime, at steady-state condition and monomolecular recombination.

By changing the intensity of the excitation one changes the density of the excited species. If bimolecular recombination is active, it will induce a shorter lifetime at higher density. Frequency-dependent plots at different excitation intensities can thus highlight this recombination mechanism. For pure monomolecular recombination, at steady state modulation and before saturation, the signal amplitude will be linear with the excitation intensity ($\tilde{N} \sim G_0$). For pure bimolecular recombinations, before saturation and steady state condition, it will be $\tilde{N} \sim \sqrt{G_0}$. Saturation always sets in at high excitation densities (see Box 9.3), masking the recombination kinetics. Finally, in real systems, there is very often a distribution of lifetimes, which can smear or completely wash out the recombination behavior.

Appendix 9.A: Pump Probe in the Two-Level System

The two-level density operator of a system with inversion symmetry so that the electric dipole operator only has off-diagonal terms μ_{eg} (no permanent dipoles) is given by the following *convolution* equations:

$$\rho_{eg}(t) = \mu_{eg} G_{eg}(t) \otimes \left\{E(t)\left[\rho_{gg}(t) - \rho_{ee}(t)\right]\right\} \tag{9.A.1a}$$

$$\rho_{ee}(t) = G_{ee}(t) \otimes \left\{E(t)\left[\mu_{eg}\rho_{ge}(t) - \rho_{eg}(t)\mu_{ge}\right]\right\} \tag{9.A.1b}$$

with impulsive response functions, $G_{eg} = \frac{i}{\hbar}\theta(t)\exp\left(-i\omega_{eg}t - t/T_2\right)$ and $G_{ee} = \frac{i}{\hbar}\theta(t)\exp\left(-t/T_1\right)$. G_{eg} is the polarization response function and G_{ee} is the population response function.

To crack the nested problem in Eq. (9.A.1), one can use the perturbation theory, provided that $\Omega_R = \frac{2\mu E_P}{\hbar} \ll \frac{1}{T_2}$. This leads to the iterative solution

$$\rho_{nm}^{(p+1)}(t) = G_{nm}(t) \otimes \left(E(t)\sum_l \left[\mu_{nl}\rho_{lm}^{(p)} - \rho_{nl}^{(p)}\mu_{lm}\right]\right) \tag{9.A.2}$$

which can be used by starting with the assumption that at $t = 0$ the whole population is in the ground state, or $\rho_{gg}^{(o)} = 1$, $\rho_{ee}^{(o)} = 0$. Inserting this ansatz in the iterative solution (Eq. (9.A.2)), the first-order term ($p = 0$) is

$$\rho_{eg}^{(1)}(t) = \mu_{eg} G_{eg}(t) \otimes \left\{E(t)\left[\rho_{gg}^{(0)} - \rho_{ee}^{(0)}\right]\right\} = \mu_{eg} G_{eg}(t) \otimes E(t) \tag{9.A.3}$$

(Remember that $\mu_{ee} = \mu_{gg} = 0$). The game goes on as long as you like, to any order, according to $\rho_{gg}^{(o)} \to \rho_{eg}^{(1)} \to \rho_{ee}^{(2)} \to \rho_{eg}^{(3)} \to \cdots$ Note that the population and polarization terms (diagonal and off-diagonal) alternate.

For $p = 2$ one gets the third-order polarization relevant to the pump probe signal:

$$\rho_{eg}^{(3)}(t) = 2\mu_{eg}\mu_{ge}\mu_{eg} G_{eg}(t) \otimes \left\{E(t)\left[G_{ee}(t) \otimes \left(E(t)\left\{\left[G_{eg}(t) + G_{ge}(t)\right] \otimes E(t)\right\}\right)\right]\right\} \tag{9.A.4}$$

This expression contains many terms, because the total electric field is the sum of the pump and the probe field, $E(\bar{r}, t) = \frac{1}{2}\left[E_P(\bar{r}, t) + E_T(\bar{r}, t) + E_P^*(\bar{r}, t) + E_T^*(\bar{r}, t)\right]$. There are 4^3 different terms, but only a few are meaningful for our experiment. The signal should propagate in the direction of the probe (K_T); thus it should contain *two pump* interactions and *one probe* interaction, and it should be resonant (energy conservation rule). By using those conditions only four terms survive. Two of them are included in the so-called population or sequential term that represents the intuitive process of pump probe as we understood it: two pump interactions generate a population that affects the probe polarization. The two other terms represent polarization coupling and pump-perturbed probe free induction decay (FID). The population term depends on the interaction sequences $E_T E_P E_P^*$ and $E_T E_P^* E_P$, both leading to a polarization with wavevector in the probe direction, for example: $\exp\left(i\bar{K}_P \cdot \bar{r}\right)\exp\left(-i\bar{K}_P \cdot \bar{r}\right)\exp\left(i\bar{K}_T \cdot \bar{r}\right) = \exp\left(i\bar{K}_T \cdot \bar{r}\right)$. The population (sequential) term is

$$\rho_{eg}^{(3)}(t) = \frac{\mu_{eg}}{2} G_{eg}(t) \otimes \left\{E_T(t)\left[\rho_{gg}^{(2,PP)}(t) - \rho_{ee}^{(2,PP)}(t)\right]\right\} \tag{9.A.5}$$

where the second-order population difference term is

$$\left[\rho_{gg}^{(2,PP)}(t) - \rho_{ee}^{(2,PP)}(t)\right] = \frac{\mu_{eg}\mu_{ge}}{2} G_{ee}(t) \otimes \left\{E_{\mathrm{P}}^{*}(t)\left[G_{eg}(t) \otimes E_{\mathrm{P}}(t)\right] - E_{\mathrm{P}}(t)\left[G_{eg}^{*}(t) \otimes E_{\mathrm{P}}^{*}(t)\right]\right\} \quad (9.A.6)$$

Note that this term depends only on the pump field.

The other two terms that contribute to $\rho_{eg}^{(3)}(t)$, and thus to the signal, are

$$\rho_{eg}^{(3,\mathrm{PTP})}(t) = \frac{\mu_{eg}\mu_{ge}\mu_{eg}}{4} G_{eg}(t) \otimes \left\{E_{\mathrm{P}}(t)\left(G_{ee}(t) \otimes \left(E_{\mathrm{T}}(t)\left[G_{eg}^{*}(t) \otimes E_{\mathrm{P}}^{*}(t)\right]\right)\right)\right\} \quad (9.A.7)$$

$$\rho_{eg}^{(3,\mathrm{PPT})}(t) = -\frac{\mu_{eg}\mu_{ge}\mu_{eg}}{4} G_{eg}(t) \otimes \left\{E_{\mathrm{P}}(t)\left(G_{ee}(t) \otimes \left(E_{\mathrm{P}}^{*}(t)\left[G_{eg}(t) \otimes E_{\mathrm{T}}(t)\right]\right)\right)\right\} \quad (9.A.8)$$

Equation (9.A.7) is the polarization coupling term due to the interaction $E_{\mathrm{P}} E_T E_{\mathrm{P}}^{*}$. The pump field generates a polarization that interferes with the probe field, forming a grating that scatters off the second pump field into the probe direction (check the wavevectors combination). This term appears only during pump probe overlap in time.

Equation (9.A.8) is the pump-perturbed probe FID due to the interaction $E_{\mathrm{P}} E_{\mathrm{P}}^{*} E_{\mathrm{T}}$. Here the probe field generates a polarization that is then detected (it is in the proper direction in space). This polarization is, however, perturbed by the pump pulse, and decays differently with respect to the no-pump situation. This leads to a pump probe signal that develops at negative times and suddenly decays after the pump pulse reached the sample.

In Figure 9.A.1, the three contributions to the pump probe signal and the total signal within the two-level approximation are shown as worked out numerically integrating the equations above. The population term is the dominant contribution at $t = 0$ and the only one at $t > 0$; however, around zero time delay and for negative time (when the probe precedes the pump pulse) the other terms are important. Pump-perturbed probe FID is also clearly present at very large negative delay and is associated with a large oscillation in spectrum. Polarization coupling is present only around zero, during pulse overlap.

Equation (9.A.5) reduces to the effective linear term Eq. (9.19) if we consider a probe pulse $E_{\mathrm{T}}(t-\tau)$that is short with respect to T_1 (the population decay time) but long with respect to T_2, the polarization (G_{eg}) decay time, and we use the property of the Fourier transform for the convolution:

$$\begin{aligned} G_{eg}\ (t) \otimes \left\{E_{\mathrm{T}}(t)\left[\rho_{gg}^{(2,\mathrm{PP})}(t) - \rho_{ee}^{(2,PP)}(t)\right]\right\} &= F\left\{G_{eg}(t)\right\} F\left\{E_{\mathrm{T}}(t-\tau)\left[\Delta N_g(t) - \Delta N_e(t)\right]\right\} \\ &= F\left\{G_{eg}(t)\right\} F\left\{E_{\mathrm{T}}(t)\right\}\left[\Delta N_g(\tau) - \Delta N_e(\tau)\right] \\ &= G_{eg}(\omega) E_{\mathrm{T}}(\omega)\left[\Delta N_g(\tau) - \Delta N_e(\tau)\right] \end{aligned} \quad (9.A.9)$$

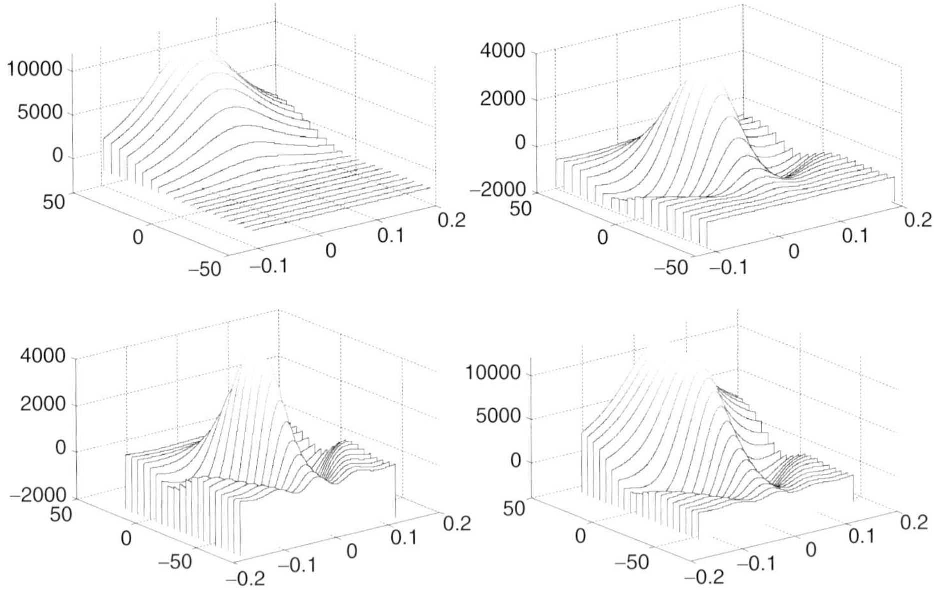

Figure 9.A.1 Simulation of pump probe contributions in the two-level model. Relevant parameters are $T_1 = 100$ fs, $T_2 = 10$ fs, $t_P = 40$ fs, $t_T = 10$ fs, and detuning $\Delta = 0$.

(where we write $\rho_{xx}^{(2,PP)}(t) = \Delta N_x(t)$). The third-order matrix element and thus the third-order polarization, is $\rho_{eg}^{(3)}(\omega) = \frac{\mu_{eg}}{2} G_{eg}(\omega) E_T(\omega) \left[\Delta N_g(\tau) - \Delta N_e(\tau)\right]$. Accordingly, the pump probe signal will be described by

$$\frac{\Delta I(\omega)}{I_0(\omega)} \approx \frac{\omega L}{\varepsilon_0 nc} \operatorname{Im} \frac{P^{(3)}(\omega)}{E_T'(\omega)} = \frac{\omega \mu_{eg}^2 NL}{\varepsilon_0 nc} \operatorname{Im} \tilde{G}(\omega) \left[\Delta N_g(\tau) - \Delta N_e(\tau)\right] \quad (9.A.10)$$

which is Eq. (9.19) if we define $\frac{\omega \mu_{eg}^2 NL}{\varepsilon_0 nc} \operatorname{Im} \tilde{G}(\omega) = \sigma(\omega) L$.

Appendix 9.B: The Coherent Coupling in Degenerate Pump Probe

We develop here a formalism to describe the pump probe interaction based on the third-order susceptibility and the slowly varying envelop approximation (SVEA). We assume that detection is spectral integrated, and the signal is the normalized change in energy of the transmitted probe. This approach provides a clear explanation for the so-called coherent artifact, observed in degenerate pump probe experiments when both pulses have the same frequency (one-color experiment).

The total electric field is

$$\overline{E}_{TOT} = \operatorname{Re}\left[\tilde{\varepsilon}_e(\vec{r}, t) e^{-i\vec{k}_e \cdot \vec{r}} + \tilde{\varepsilon}_p(\vec{r}, t - \tau) e^{-i\vec{k}_p \cdot \vec{r}}\right] \quad (9.B.11)$$

where the time-dependent field amplitude is $\tilde{\varepsilon}(\vec{r},t) = \overline{E}(\vec{r},t)\,e^{i\omega t}$. The polarization in the medium is described similarly as

$$\overline{P} = \mathrm{Re}\left[\tilde{p}(\vec{r},t)e^{-i\bar{k}\cdot\bar{r}}\right] \tag{9.B.12}$$

with

$$\tilde{p}(\vec{r},t) = \varepsilon_0\,(\tilde{\chi}_\mathrm{L} + \tilde{\chi}_\mathrm{NL})\,\tilde{\varepsilon}(\vec{r},t) = \tilde{p}_\mathrm{L} + \tilde{p}_\mathrm{NL} \tag{9.B.13}$$

The wave equation under SVEA approximation $\left(\frac{\partial^2\tilde{\varepsilon}}{\partial z^2} \ll k\frac{\partial\tilde{\varepsilon}}{\partial z} \text{ and } \frac{\partial\tilde{\varepsilon}}{\partial t} = i\omega\mathrm{E}\right)$ for linear polarized fields in one direction is

$$k\frac{\partial\tilde{\varepsilon}}{\partial z} = \frac{\omega^2}{2c^2}\chi_\mathrm{L}''\tilde{\varepsilon} - i\frac{\omega^2}{2c^2\varepsilon_0}\tilde{p}_\mathrm{NL}e^{ikz} \tag{9.B.14}$$

where time derivatives are assumed to be negligible during the transit time $t = \frac{l\sqrt{\varepsilon}}{c}$ through the sample of thickness l. We see from the wave equation that the nonlinear polarization produces a contribution proportional to ip_NL. If p_NL is pure real quantity, this will cause a phase shift in the incoming field. If p_NL is pure imaginary, it will cause a change in amplitude of the incoming field. Now we assume that the electronic dephasing time $T_2 \to 0$, that is, the polarization reacts instantaneously to the applied field. Under this assumption we can disregard the nonlocal response and write $P^{(3)}(t) = \tilde{\chi}_\mathrm{NL}(t)E(t)$, where we drop the vector notation. In general, $\tilde{\chi}_\mathrm{NL} = \tilde{\chi}_\mathrm{NL}' + \tilde{\chi}_\mathrm{NL}''$ is a complex tensor. We drop the tensor notation and the vector notation. The imaginary part of the nonlinear susceptibility is given by the convolution of the *impulsive* material response function with the square total field: $\tilde{\chi}_{\mathrm{NL},a}'' = \int_{-\infty}^{t} A(t-t')E_\mathrm{TOT}(t')E_\mathrm{TOT}^*(t')\mathrm{d}t'$. There are four terms and their complex conjugates: $\left|\tilde{\varepsilon}_e(\vec{r},t)\right|^2$ $\quad\tilde{\varepsilon}_e^*(\vec{r},t)\tilde{\varepsilon}_p(\vec{r},t-\tau)e^{i(\bar{k}_e-\bar{k}_p)\cdot\bar{r}}$ $\quad\tilde{\varepsilon}_e(\vec{r},t)\tilde{\varepsilon}_p^*(\vec{r},t-\tau)e^{i(\bar{k}_p-\bar{k}_e)\cdot\bar{r}}$ $\quad\left|\tilde{\varepsilon}_p(\vec{r},t)\right|^2$

The last one is weak, according to the pump probe approximation, and can be disregarded.

The first gives rise to the "*population*" term:

$$\tilde{\chi}_{\mathrm{NL},a}'' = \int_{-\infty}^{t} A(t-t')\left|\varepsilon_e(t')\right|^2\mathrm{d}t' \tag{9.B.15}$$

The second and third give rise to the grating term:

$$\tilde{\chi}_{\mathrm{NL,g}}' = e^{-i(\bar{k}_p-\bar{k}_e)\cdot\bar{r}}\int_{-\infty}^{t} A(t-t')\varepsilon_e^*(t')\varepsilon_p(t'-\tau)\mathrm{d}t' + c.c. \tag{9.B.16}$$

This term is due to the interference of the pump and probe field in the material, generating a grating of wavevector $\bar{k}_g = \bar{k}_p - \bar{k}_e$. Note that for nondegenerate experiments with $\omega_e \neq \omega_p$ this term is not important because the interference is washed out in time (averaging $\varepsilon_e^*\varepsilon_p \propto e^{i(\omega_p-\omega_e)t'}$).

The resulting polarization in the sample, according to our approximation of instantaneous dephasing, is given by $\tilde{p}_\mathrm{NL} = \varepsilon_0\chi_\mathrm{NL}(t)\tilde{\varepsilon}_\mathrm{T}(t)$ with $\tilde{\varepsilon}_\mathrm{T} = \tilde{\varepsilon}_e(t)e^{-i\bar{k}_e\cdot\bar{r}} + \tilde{\varepsilon}_p(t-\tau)e^{-i\bar{k}_p\cdot\bar{r}}$. By expanding the product we have several terms. The pump-only contribution $\propto \tilde{\varepsilon}_e^*(\vec{r},t)\left|\tilde{\varepsilon}_e(\vec{r},t)\right|^2$ is a self-action of the pump on itself, it is in the pump direction k_e and is not measured.

The next terms in the probe direction k_p are

$$\tilde{p}_{\mathrm{NL},a} = i\varepsilon_o\tilde{\varepsilon}_p(t-\tau)e^{-i\bar{k}_p\cdot\bar{r}}\int_{-\infty}^{t} A(t-t')\left|\varepsilon_e(t')\right|^2 dt' \tag{9.B.17}$$

$$\tilde{p}_{\mathrm{NL},g} = i\varepsilon_o\tilde{\varepsilon}_e(t)e^{-i\bar{k}_p\cdot\bar{r}}\int_{-\infty}^{t} A(t-t')\varepsilon_p(t'-\tau)\varepsilon_e^*(t')dt' \tag{9.B.18}$$

Finally, there is a contribution in a new direction $\bar{k}_S = 2\bar{k}_e - \bar{k}_p$

$$\tilde{p}_{\mathrm{NL},S} = i\varepsilon_o\tilde{\varepsilon}_e^*(t)e^{i\bar{k}_S\cdot\bar{r}}\int_{-\infty}^{t} A(t-t')\varepsilon_p(t'-\tau)\varepsilon_e^*(t')dt' \tag{9.B.19}$$

The last two terms are due to the pump field diffraction off the interference grating. $\tilde{p}_{\mathrm{NL},a}$ is in the probe direction, it will contribute to the pump probe signal. $\tilde{p}_{\mathrm{NL},S}$ is in the direction $\bar{k}_S = 2\bar{k}_e - \bar{k}_p$. It is called *two pulse scattering*.

The SVEA wave equation along k_p is

$$\frac{\partial\tilde{\varepsilon}_p}{\partial z} = -\frac{\alpha_0}{2}\tilde{\varepsilon}_p + \frac{\omega^2}{2kc^2}\left[Q_1(t) + Q_2(t)\right]e^{-3\frac{\alpha_0}{2}z} \tag{9.B.20}$$

where we used $\alpha_0 = \frac{\omega^2}{kc^2}\chi_{\mathrm{L}}''$

$$Q_1(t,\tau) = \tilde{\varepsilon}_p(t-\tau)\int_{-\infty}^{t} A(t-t')\left|\varepsilon_e(t')\right|^2 dt' \tag{9.B.21}$$

$$Q_2(t,\tau) = \tilde{\varepsilon}_e(t)\int_{-\infty}^{t} A(t-t')\varepsilon_p(t'-\tau)\varepsilon_e^*(t')dt' \tag{9.B.22}$$

The integrated field is

$$\tilde{\varepsilon}_p(z,t) = e^{-\frac{\alpha_0}{2}L}\left\{\tilde{\varepsilon}_p(0,t) + \frac{\omega^2}{2kc^2}\left[Q_1(t,\tau) + Q_2(t,\tau)\right]\eta\right\} \tag{9.B.23}$$

with $\eta = \frac{1-e^{-\alpha_0 L}}{\alpha_0}$. The input intensity is $I_{\mathrm{in}} \cong \left|\tilde{\varepsilon}_p(0,t-\tau)\right|^2$, while the transmitted intensity can be written as a linear and nonlinear contribution:

$$I_{\mathrm{out}}^{\mathrm{L}} \cong \left|\tilde{\varepsilon}_p(0,t-\tau)\right|^2 e^{-\alpha_0 L} \tag{9.B.24}$$

$$\begin{aligned} I_{\mathrm{out}}^{\mathrm{NL}} &\propto \tilde{\varepsilon}_p\tilde{\varepsilon}_p^* \\ &\cong \left\{\left|\tilde{\varepsilon}_p(0,t-\tau)\right|^2 + \frac{\omega^2}{2kc^2}\tilde{\varepsilon}_p(0,t-\tau)\left[Q_1(t,\tau) + Q_2(t,\tau)\right]\eta + c.c.\right\} \\ &\times e^{-\alpha_0 L} \end{aligned} \tag{9.B.25}$$

where we disregard terms in $\left|Q_1(t,\tau) + Q_2(t,\tau)\right|^2$.

Finally, the normalized transmission change is

$$\frac{\Delta I(t)}{I_{\mathrm{in}}} = \frac{\left\{\frac{\omega^2\eta}{2kc^2}\tilde{\varepsilon}_p(0,t-\tau)\left[Q_1(t,\tau) + Q_2(t,\tau)\right] + c.c.\right\} \times e^{-\alpha_0 L}}{\tilde{\varepsilon}_p(0,t)} \tag{9.B.26}$$

Let us assume here that the detector makes the time integral of this signal, on time $t_D \gg 2\pi/\omega$.

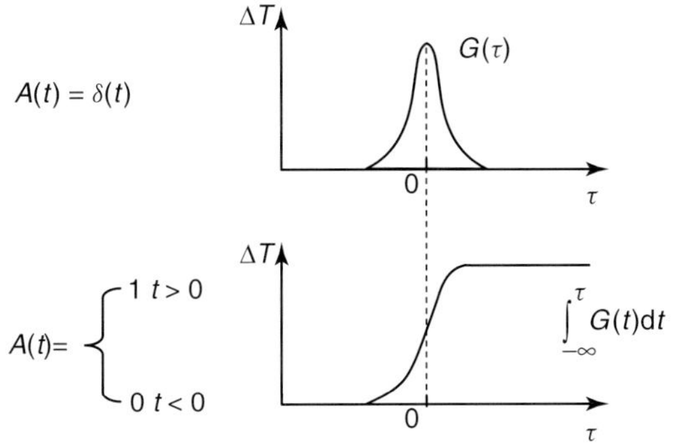

Figure 9.B.2 The population term in a pump probe experiment is the convolution of the pulse autocorrelation $G(t)$ with the impulsive response function $A(t)$. Two extreme cases are depicted.

The time behavior of the two contributions can be evaluated as follows. For Q_1,

$$\frac{\Delta I(\tau)}{I_{\text{in}}} = \int \frac{\Delta I(t)}{I_{\text{in}}} \mathrm{d}t = \int \tilde{\varepsilon}_p(t-\tau)\tilde{\varepsilon}_p^*(t-\tau)\mathrm{d}t \int_{-\infty}^{t} A(t-t')\left|\varepsilon_e(t')\right|^2 \mathrm{d}t' \tag{9.B.27}$$

which is the convolution of the pulse cross-correlation with the response function[1)]

$$\frac{\Delta I(\tau)}{I_{\text{in}}} = \int_{-\infty}^{\tau} A(\tau - t')G(t')\mathrm{d}t' \tag{9.B.28}$$

where $G(t) = \int \left|\tilde{\varepsilon}_p(t'-t)\right|^2 \left|\varepsilon_e(t')\right|^2 \mathrm{d}t'$ is the intensity pulse autocorrelation. This tells us that the resolution in time depends on the pulse autocorrelation. In Figure 9.B.2 we consider the two limiting cases. If $A(t)$ is a δ-function with respect to $G(t)$, the recorded signal will be $G(t)$. Any information on the material dynamics is thus lost. If $A(t)$ is a step function, the recorded signal is a step, with a rising edge following $G(t)$ and a plateau.

The time behavior of Q_2 is

$$\frac{\Delta I(\tau)}{I_{\text{in}}} = \int \frac{\Delta I(t)}{I_{\text{in}}} \mathrm{d}t = \int \tilde{\varepsilon}_p(t-\tau)\tilde{\varepsilon}_e(t) \int_{-\infty}^{t} A(t-t')\varepsilon_p^*(t'-\tau)\varepsilon_e(t')\mathrm{d}t' \tag{9.B.29}$$

Here if $A(t)$ is a δ-function with respect to $G(t)$ we obtain $G(t)$ again. If $A(t)$ is a step function the contribution to the signal is the square modulus of the electric field autocorrelation, $\left|\int_{-\infty}^{+\infty} \varepsilon_p(t-\tau)\varepsilon_e(t)\mathrm{d}t\right|^2$. In both situations, this contribution is

1) $\tilde{F}(I_p{}^*(A \times I_e)) = \tilde{F}^*(I_p)\tilde{F}(A \times I_e) = \tilde{F}^*(I_p)\tilde{F}(A)\tilde{F}(I_e) = \tilde{F}(A)\left[\tilde{F}^*(I_p)\tilde{F}(I_e)\right] = \tilde{F}^*(A)\tilde{F}(I_p{}^* I_e) = A \times (I_p{}^* I_e)$.

a symmetric peak around $t = 0$, which does not bring information on the sample dynamics. For this reason, this contribution is called *coherent coupling* or *artifact*. In two-color experiments with narrow band pulses this signal does not appear, because pulse interference is quickly washed out in time. In one-color experiments or when pump and probe are broadband and have some spectral overlap, this contribution adds to the population signal. Because it is confined to the pulse overlap time domain this does not affect the longtime signal. Only when the very initial dynamics is of concern, and the sample response time is comparable to the pulse width, this contribution should be carefully evaluated.

In summary, according to this theory and the assumption that $T_2 \rightarrow 0$, the measured signal is the sum of two contributions, the population term and the coherent coupling term. The latter extends symmetrically around $t = 0$ and has a width of the order of the pulse correlation.

So far we disregarded the electric field vector components, but they play a role in the relative intensity of the signal. For pump and probe with parallel polarization, let us say "z," the involved third-order tensor component, is χ^3_{zzzz} for both population and coherent coupling signal. In this configuration and at zero pump probe delay the amplitude of the coherent coupling is exactly half the total signal. For perpendicularly polarized pump and probe the population term is $\propto \chi^3_{xxzz}$, while the coherent coupling term $\propto \chi^3_{xzzx}$. If the sample is isotropic and the polarization memory is lost instantaneously the coherent coupling will not appear in the signal. If the sample is isotropic and the polarization memory retained, the coherent coupling will have an amplitude, at $t = 0$, between 0 and 1/3 of the population term.

10
Conclusion

The goal of this book is to provide an easy access guide to photophysics. It contains knowledge that stays behind applications in photovoltaics, optoelectronics, and photonics. While such applications will keep changing and technology evolution will never stop, its basic will stay unchanged. Indeed, many of the topics discussed here rest on last century physics or chemistry, while modern evolution mainly concerns materials, not really phenomena. However, what will change is the interrelation between disciplines.

Advanced systems based on nanotechnology evolve toward the integration with biology. For instance, in this book we mention biomimetic and bioinspired approaches to light conversion. More comes from the blowing out of humanoid technology, as in robotics and medicine, or life-enhancing technology and genetics. There are huge challenges for materials and devices in such areas of application, which include sensors and detectors, information processing and transfer, control motion and positioning, and more. The new researcher should have a broad view and be multidisciplinary. However, we cannot go back to the Renaissance man of science and art, who was all in one, for the amount of knowledge developed nowadays is huge and above any human capability. I believe the general principle in research be *strengthens the strengths*, which is the basic concept for working in a team where each player has a role. So how to put together the two contradicting issues, multidisciplinarity and specialization? I think one needs to have a clear "major" on top of a broad background. The major is one's own job, that is, the field where one has professional skills, and that provides the means for contributing to realization of ideas. This is the role in the team, when the game gets going. The broad background is a cultural tool, for contributing with ideas to the overall strategy of the project and for having an intercommunication language.

Science should not stay in an ivory tower, but always be at the service of mankind, which means science should be applied to become technology. I believe this is the most exciting challenge in research, and for many practical reasons the one dominating it. The technology we had in mind here is photovoltaic conversion or any other practical use of light, that is, photonics. What are the directions of development in these technologies?

The Photophysics behind Photovoltaics and Photonics, First Edition. Guglielmo Lanzani.

Any prediction of the future, in whatsoever field, is a personal view with no scientific authority, from whomever it comes. Keeping this in mind, one can still make projections as extensions of the present trends.

In our field, I think evolution will bring about

1) new materials
2) better system integration
3) new phenomena.

Point 1 is the obvious quest for better materials in terms of known required properties: better absorption, better transport, better stability, better nonlinear response, and so on. These are requirements for improving existing devices while preserving their shape and working mechanisms, in other words, replacing one or more components into a device, to improve its performance.

Point 2 means to add systems up. In photovoltaics, it could be a light collector device coupled to the photovoltaic cell, to enhance the cross section of light collection, or, for instance, for spectral conversion of the light before conversion to energy. It can also be tandem cells, the idea to couple devices optimized for different spectral regions. It can be device integration such as a photovoltaic cell running a water-splitting device to finally convert sunlight into hydrogen, or an electrochromic window. In photonics, it can be the integration of organic materials with silicon, or plasmonic nanoparticles into switches or passive optical components.

Point 3 is the use of different phenomena, such as carrier multiplication, or singlet fission, or hot charge separation, or the emergent concept of storage of energy in forms other than a separated charge pair, such as molecular conformation. Here we can mention the growing effort toward water splitting, aiming at using sunlight to produce H and O to be recombined in fuel cells. This research is strongly bioinspired and is a good example of multidisciplinarity in action. In photonics, many different phenomena are proposed for realizing time domain switching, where some of the concepts we discussed are exploited for redirecting, enhancing, or reducing light.

Innovation by devising new phenomena is an exciting activity that stems from creativity. Surprises sure will come. Ideas will sprout whereever a good enough scientific background is available to conceive that, and good enough technology is available for testing the new ideas. Successful ideas should indeed come at the right time, when the right technology is available. Suppose you can travel in time. You could go to the future, take some idea, come back, and realize it before others, getting famous and rich. Careful, however, if you go too much ahead in time, you will not be able to import anything suitable because the present technology will not allow you to realize the future idea. Timing is essential and sometimes punitive for the creativeness of people. You might have a genial idea at the wrong time, and never see it sprouting. Leonardo da Vinci would certainly have realized a working hang glider, or maybe an airplane, having the right materials and technology.

Basic science should never be lost, forgot, or left apart; it is and it always will be at the heart of development. This is true even in a fast changing world dominated by High Tech, even where professions keep changing names and go toward a capillary specialization. We are talking about creativity, new ideas, and their realization, new frontiers where there are no roads and no maps. There what you need is solid equipment, made of traditional gears that in a way or the other will help you make the trail.

Bibliography

Archipov, V.I., Fishchuk, I.I., kadashchuck, A., and Baessler, H. (2006) Charge transport in disordered organic semiconductors, in *Photophysics of Molecular Materials*, Chapter 6 (ed. G. Lanzani), Wiley-VCH Verlag GmbH, KGaA, Weinheim, pp. 261–358. ISBN: 3-527-40456-2.

Atkins, P. and Friedman, R. (2005) *Molecular Quantum Mechanics*, Oxford University Press, New York. ISBN: 0-19-927498-3.

Braun, C.L. (1984) *J. Chem. Phys.*, **80**, 4157.

Cabanillas-Gonzalez, J. *et al.* (2006) *Phys. Rev. Lett.*, **96**, 106601.

Deisenhofer, J. *et al.* (1984) *J. Mol. Biol.*, **180**, 385–398.

Dexter, D.L. (1953) *J Chem. Phys.*, **21**, 836.

IRAV in polymers are described in Eherenfreund, E., Vardeny, Z.V., Brafman, O., and Horovitz, B. (1987) *Phys. Rev.*, **B36**, 1535.

Engel, G.S., Calhoun, T.R., Read, E.L., Ahn, T.-K., Mancal, T., Cheng, Y.-C., Blankenship, R.E., and Fleming, G.R. (2007) *Nature*, **446**, 782–786.

(a) Foerster, T. (1948) *Ann. Phys. (Leipzig)*, **2**, 55; (b) Foerster, T. (1959) *Discuss. Faraday Soc.*, **27**, 7.

Gadermaier, C. *et al.* (2002) *Phys. Rev. Lett*, **89**, 117402.

Gambetta, A., Manzoni, C., Menna, E., Meneghetti, M., Cerullo, G., Lanzani, G., Tretiak S., Piryatinski, A., Saxena, A., Martin R.L., and Bishop, A.R. (2006) *Nat. Phys.*, **2**, 515–520.

Garbugli, M. *et al.* (2009) *J. Mater. Chem.*, **19**, 7551–7560. (Evidence for the essential state model and charge generation).

Greene, B.I., Orenstein, J., and Schmitt-Rink, S. (1990) *Science*, **247**, 679–687.

Gulbinas, V. *et al.* (2002) *Phys. Rev. B*, **66**, 233203.

Hoffert, M.I. *et al.* (2002) *Science*, **298**, 981–987.

Hudson, B.S., Kohler, B.E., and Schulten, K. (1982) *Excited Stares*, Vol. 6, Academic Press, New York, p. 1.

Ishizaki, A., Calhoun, T.R., Schlau-Cohen, G.S., and Fleming, G.R. (2010) *Phys. Chem. Chem. Phys.*, **12**, 7319–7337.

Ishizaki, A. and Fleming, G.R. (2009a) *J. Chem. Phys.*, **130**, 234111.

Ishizaki, A. and Fleming, G.R. (2009b) *Proc. Natl. Acad. Sci. U.S.A.*, **106**, 17255–17260.

Joffre, M. (1988) Coherent effects in femtosecond spectroscopy. A simple picture using the Bloch equations, in *Femtosecond Laser Pulses* (ed. C. Rulliere), Springer, Berlin, pp. 261–283.

Kasha, M. (1950) *Discuss. Faraday Soc.*, **9**, 14.

Kittel, C. (1995) *Introduction to Solid State Physics*, Wiley-VCH Verlag GmbH. ISBN: 9971-51-180-0.

Krahne, R., Morello, G., Figuerola, A., George, C., Deka, S., and Manna, L. (2011) Physical properties of elongated inorganic nanoparticles. *Phys. Rep.-Rev. Sect. Phys. Lett.*, **501** (3–5), 75–221. DOI: 10.1016/j.physrep.2011.01.001.

Lanzani, G. (ed.) (2006) *Photophysics of Molecular Materials*, Wiley-VCH Verlag GmbH, Weinheim. ISBN: 3-527-40456-2.

Lanzani, G. *et al.* (2001) *Phys. Rev. Lett.*, **87**, 187402.

Laubereau, A. and Kaiser, W. (1978) *Rev. Mod. Phys.*, **50**, 607.

Lüer, L., Hoseinkhani, S., Polli, D., Crochet, J., Hertel, T., and Lanzani, G. (2008) *Nat. Phys.*, **5**, 54–58. ISSN: 1745-2473.

The Photophysics behind Photovoltaics and Photonics, First Edition. Guglielmo Lanzani.

Marcus, R. (1993) *Rev. Mod. Phys.*, **65**, 599.
Novoderezhkin, V.I. *et al.* (2004) *J. Phys. Chem. B*, **108**, 7445–7457.
Novoderezhkin, V.I., Dekker, J.P., and van Grondelle, R. (2007) *Biophys. J.*, **93**, 1293–1311.
Onsager, L. (1938) *Phys. Rev.*, **54**, 554.
Oppenheimer, J.R. (1941) *Phys. Rev.*, **60**, 158.
Pai, D.M. and Enck, R.C. (1975) *Phys. Rev.*, **B11**, 5163.
Prezhdo, O.V. (2008) *Chem. Phys. Lett.*, **460**, 1–9.
Sarovar, M., Ishizaki, A., Fleming, G.R., and Whaley, K.B. (2010) *Nat. Phys.*, **5**, 1.
Siebrand, W. (1966) *J. Chem. Phys.*, **44**, 4055.
Silinsh, E.A. and Inokuchi, H. (1991) *Chem. Phys.*, **149**, 373.
Spano, F. (2005) *J. Chem. Phys.*, **122**, 234701.
Tong, M. *et al.* (2007) *Phys. Rev.*, **B75**, 125207. (Evidence for the essential state model based on non linear spectroscopy).
Virgili, T., Luer, L., Cerullo, G., Lanzani, G., Stagira, S., Coles, D., Meijer, A.J.H.M., and Lidzey, D.G. (2010) *Phys. Rev. B*, **81**, 125317.
Volrkurn, R.H. and Michel-Beyerle, M.E. (1973) *Chem. Phys. Lett.*, **23**, 128.
Wang, F., Dukovic, G., Brus, L.E., and Heinz, T.F. (2005) *Science*, **308**, 838–841.
Yakovlev, A.G., Shkuropatov, A.Y., and Shuvalov, V.A. (2002) *Biochemistry*, **41**, 14019–14027.
Yariv, A. (1988) *Quatum Electronics*, Wiley-VCH Verlag GmbH, New York. ISBN: 0-471-60997-8.

Some References on Conjugated Polymers

Archipov, V.I. *et al.* (1995) *Phys. Rev. B*, **52**, 4932.
Archipov, V.I. *et al.* (1999) *Phys. Rev. Lett.*, **82**, 1321.
Graupner, W. *et al.* (1998) *Phys. Rev. Lett.*, **81**, 3259.
Heeger, A.J. *et al.* (1988) *Rev. Mod. Phys.*, **60**, 781.
Infelta, P.R., de Haas, M.P., and Warman, J.M. (1977) *Radiat. Phys. Chem.*, **10**, 353.
Kohler, A. *et al.* (1998) *Nature*, **392**, 903.
Virgili, T. *et al.* (2005) *Phys. Rev. Lett.*, **94**, 117402.
Vissenberg, M.C.J. *et al.* (1996) *Phys. Rev. Lett.*, **77**, 4820.
Wohlgenannt, M. *et al.* (1999) *Phys. Rev. Lett.*, **82**, 3344.

Index

The Photophysics behind Photovoltaics and Photonics, First Edition. Guglielmo Lanzani.